ASVAB Math Formula Sheet and Key Points

Quick Study Guide and Test Prep Book for Beginners and Advanced Students + Two ASVAB Math Practice Tests

Dr. Abolfazl Nazari

Copyright © 2024 Dr. Abolfazl Nazari

PUBLISHED BY EFFORTLESS MATH EDUCATION

EFFORTLESSMATH.COM

All rights reserved. No part of this publication may be reproduced, distributed, or transmitted in any form or by any means, including photocopying, recording, or other electronic or mechanical methods, without the prior written permission of the author, except in the case of brief quotations embodied in critical reviews and certain other noncommercial uses permitted by copyright law, including Section 107 or 108 of the 1976 United States Copyright Act.

Copyright ©2024

ASVAB Math Formula Sheet and Key Points

2024

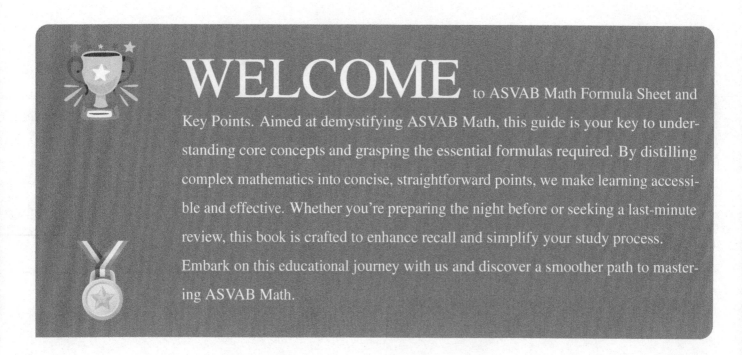

WELCOME to ASVAB Math Formula Sheet and Key Points. Aimed at demystifying ASVAB Math, this guide is your key to understanding core concepts and grasping the essential formulas required. By distilling complex mathematics into concise, straightforward points, we make learning accessible and effective. Whether you're preparing the night before or seeking a last-minute review, this book is crafted to enhance recall and simplify your study process. Embark on this educational journey with us and discover a smoother path to mastering ASVAB Math.

ASVAB Math Formula Sheet and Key Points is for students that are eager to learn ASVAB Math essentials in the fastest way. It distills complex topics down to bite-sized, manageable key points, alongside must-know formulas, facilitating quick learning and retention.

Experience a groundbreaking approach to math learning, designed for quick comprehension and lasting memory of crucial concepts. Understanding the challenge of locating all necessary formulas in one spot, we've dedicated a chapter to compiling every formula vital for the ASVAB Math examination.

What is included in this book

- ☑ Additional online resources for extended practice and support.
- ☑ A short manual on how to use this book.
- ☑ Coverage of all ASVAB Math subjects and topics tested.
- ☑ A dedicated chapter listing all the necessary formulas.
- ☑ Strategies for approaching the ASVAB Math test.
- ☑ Two comprehensive, full-length practice test with thorough answer explanations.

Effortless Math's ASVAB Online Center

Effortless Math Online ASVAB Center offers a complete study program, including the following:

- ☑ *Step-by-step instructions on how to prepare for the ASVAB Math test*
- ☑ *Numerous ASVAB Math worksheets to help you measure your math skills*
- ☑ *Complete list of ASVAB Math formulas*
- ☑ *Video lessons for all ASVAB Math topics*
- ☑ *Full-length ASVAB Math practice tests*

Visit EffortlessMath.com/ASVAB to find your online ASVAB Math resources.

Scan this QR code

(No Registration Required)

Tips for Making the Most of This Book

This book is your fastest route to mastering ASVAB Math. It distills the subject down to key points and includes the formulas you need to remember for the ASVAB Math test. Here's what makes it special:

- Concise content that focuses on the essentials.
- A dedicated chapter for formulas to remember, covering all topics.
- Two practice tests at the end of the book to test your knowledge and readiness.

Mathematics can be approachable and easy with the right tools and mindset. Our goal is to simplify math for you, focusing on what's necessary. Using the key points and the formula sheets, you can quickly review and remember the most important information.

Here's how to leverage this book effectively:

Understand and Remember Key Points

Each math topic is built around core ideas or concepts. We've highlighted key points in every topic as mini-summaries of critical information. Don't skip these!

Formulas to Remember

At the end of each topic, we've included a list of formulas to remember. These are the essential formulas you need to know for the ASVAB Math test. Make sure to review these regularly.

In summary:

- *Key Points*: Essential summaries of major concepts.
- *Formulas to Remember*: Each topic ends with a list of formulas to remember.

Once you have finished all chapters, we recommend that you review the formula sheet before taking the practice tests. This will help you recall and apply the formulas effectively.

The Practice Tests

The practice tests at the end of the book are designed to simulate the actual ASVAB Math test. They are a great way to gauge your readiness and identify areas that need more attention. Make sure to time yourself and simulate the actual test environment as closely as possible. Once you've completed the practice tests, review your answers and identify areas that need more practice.

Effective Test Preparation

A solid test preparation plan is key. Beyond understanding concepts, strategic study and practice under exam conditions are vital.

- **Begin Early**: Start studying well in advance.
- **Daily Study Sessions**: Regular, short study periods enhance retention.
- **Active Note-Taking**: Helps internalize concepts and improves focus.
- **Review Challenges**: Focus extra time on difficult topics.
- **Practice**: Use end-of-chapter problems and additional resources for extensive practice.

Pick the right study environment and materials

In addition to this book, you may want to use other resources to help you prepare for the ASVAB Math test. Here are some suggestions: If you need more explanations for the topics, you can use the *ASVAB Math Made Easy* study guide. If you need more practice, you can use our *ASVAB Math Workbook*. If you need more practice tests, you can use our *5 Full Length Test for ASVAB Math* book.

Contents

1 Fractions and Mixed Numbers 1
1.1   Simplifying Fractions 1
1.2   Adding and Subtracting Fractions 2
1.3   Multiplying and Dividing Fractions 2
1.4   Adding Mixed Numbers 3
1.5   Subtracting Mixed Numbers 4
1.6   Multiplying and Dividing Mixed Numbers 4

2 Decimals 5
2.1   Comparing Decimals 5
2.2   Rounding Decimals 5
2.3   Adding and Subtracting Decimals 5
2.4   Multiplying and Dividing Decimals 6

3 Integers and Order of Operations 7
3.1   Adding and Subtracting Integers 7
3.2   Multiplying and Dividing Integers 8
3.3   Order of Operations 8

3.4	Absolute Value	8

4 Ratios and Proportions . . . 10

4.1	Simplifying Ratios	10
4.2	Proportional Ratios	10
4.3	Similarity and Ratios	11

5 Percentage . . . 12

5.1	Percent Problems	12
5.2	Percent of Increase and Decrease	13
5.3	Discount, Tax, and Tip	13
5.4	Simple Interest	14

6 Exponents and Variables . . . 15

6.1	Multiplication Property of Exponents	15
6.2	Division Property of Exponents	16
6.3	Zero and Negative Exponents	16
6.4	Working with Negative Bases	17
6.5	Scientific Notation	17
6.6	Radicals	18

7 Expressions and Variables . . . 19

7.1	Simplifying Algebraic Expressions	19
7.2	Simplifying Polynomial Expressions	19
7.3	The Distributive Property	20
7.4	Evaluating One Variable	20
7.5	Evaluating Two Variables	20

8 Equations and Inequalities 21
8.1 One-Step Equations 21
8.2 Multi-Step Equations 22
8.3 System of Equations 22
8.4 One–Step Inequalities 22
8.5 Multi-Step Inequalities 23
8.6 Graphing Single-Variable Inequalities 23

9 Lines and Slope 24
9.1 Finding Slope 24
9.2 Graphing Lines Using Slope-Intercept Form 24
9.3 Writing Linear Equations 25
9.4 Finding Midpoint 25
9.5 Finding Distance of Two Points 25
9.6 Graphing Linear Inequalities 26

10 Polynomials 27
10.1 Simplifying Polynomials 27
10.2 Adding and Subtracting Polynomials 27
10.3 Multiplying and Dividing Monomials 28
10.4 Multiplying a Polynomial and a Monomial 28
10.5 Multiplying Binomials 28
10.6 Factoring Trinomials 29

11 Geometry and Solid Figures 30
11.1 Complementary and Supplementary angles 30
11.2 Parallel Lines and Transversals 30

11.3	Triangles	31
11.4	The Pythagorean Theorem	31
11.5	Special Right Triangles	31
11.6	Polygons	32
11.7	Circles	33
11.8	Cubes	33
11.9	Rectangular Prisms	34
11.10	Cylinder	34
12	**Statistics**	**36**
12.1	Mean, Median, Mode, and Range of the Given Data	36
12.2	Pie Graph	37
12.3	Probability Problems	37
12.4	Permutations and Combinations	38
13	**Functions Operations**	**40**
13.1	Function Notation and Evaluation	40
13.2	Adding and Subtracting Functions	40
13.3	Multiplying and Dividing Functions	41
13.4	Composition of Functions	41
14	**ASVAB Test Review and Strategies**	**42**
14.1	The ASVAB Test Review and Scoring	42
14.2	ASVAB Math Test-Taking Strategies	43
15	**Formula Sheet for All Topics**	**46**

16 Practice Test 1 58
16.1 Practices 58
16.2 Answer Keys 71
16.3 Answers with Explanation 73

17 Practice Test 2 84
17.1 Practices 84
17.2 Answer Keys 97
17.3 Answers with Explanation 99

1. Fractions and Mixed Numbers

1.1 Simplifying Fractions

Key Point

A fraction represents a part of a whole. The top number, or numerator, shows how many parts we have. The bottom number, or denominator, indicates how many parts make up the whole.

Key Point

To simplify a fraction, either: 1) Repeatedly divide the numerator and denominator by numbers such as 2, 3, 4, ... or 2) find and divide the numerator and denominator by their greatest common divisor (GCD). This single step brings the fraction to its simplest form, where no further division is possible except by 1.

Formula To Remember

1. *Simplification by Common Divisor* 👉 $\frac{a}{b} = \frac{a \div d}{b \div d} = \frac{a'}{b'}$, where $d > 1$ is a common divisor of a and b.
2. *Simplification using GCD* 👉 $\frac{a}{b} = \frac{a \div \gcd(a,b)}{b \div \gcd(a,b)} = \frac{a''}{b''}$, where $\gcd(a,b)$ is the greatest common divisor of a and b.

1.2 Adding and Subtracting Fractions

🔔 Key Point

When adding or subtracting 'like' fractions, simply add or subtract the numerators and write the result over the common denominator. So,

$$\frac{a}{b} + \frac{c}{b} = \frac{a+c}{b} \quad \text{and} \quad \frac{a}{b} - \frac{c}{b} = \frac{a-c}{b}.$$

🔔 Key Point

When adding or subtracting 'unlike' fractions, apply the following formulas:

$$\frac{a}{b} + \frac{c}{d} = \frac{ad+bc}{bd} \quad \text{and} \quad \frac{a}{b} - \frac{c}{d} = \frac{ad-bc}{bd}$$

🔔 Key Point

When adding or subtracting fractions that have different denominators, find the Least Common Denominator (LCD). The LCD is the smallest number that all the denominators can divide into without a remainder.

📌 Formula To Remember

1. Adding 'Like' Fractions 👉 $\frac{a}{b} + \frac{c}{b} = \frac{a+c}{b}$
2. Subtracting 'Like' Fractions 👉 $\frac{a}{b} - \frac{c}{b} = \frac{a-c}{b}$
3. Adding 'Unlike' Fractions 👉 $\frac{a}{b} + \frac{c}{d} = \frac{ad+bc}{bd}$
4. Subtracting 'Unlike' Fractions 👉 $\frac{a}{b} - \frac{c}{d} = \frac{ad-bc}{bd}$

1.3 Multiplying and Dividing Fractions

🔔 Key Point

To multiply fractions, simply multiply the numerators to form the new numerator, and multiply the denominators to form the new denominator. All equations can be represented as:

$$\frac{a}{b} \times \frac{c}{d} = \frac{a \times c}{b \times d}.$$

1.4 Adding Mixed Numbers

Key Point

To divide fractions, keep the first fraction the same, change the division sign into multiplication, and flip the second fraction (called the reciprocal). The structure then becomes:

$$\frac{a}{b} \div \frac{c}{d} = \frac{a}{b} \times \frac{d}{c} = \frac{a \times d}{b \times c}.$$

Formula To Remember

1. *Multiplication of Fractions* ☞ $\frac{a}{b} \times \frac{c}{d} = \frac{a \times c}{b \times d}$.
2. *Division of Fractions (Keep, Change, Flip)* ☞ $\frac{a}{b} \div \frac{c}{d} = \frac{a}{b} \times \frac{d}{c} = \frac{a \times d}{b \times c}$.

1.4 Adding Mixed Numbers

Key Point

To add mixed numbers, start by splitting them into whole numbers and fractions. Add the whole numbers together. Then, add the fractions. Finally, combine these sums into a new mixed number.

Key Point

To convert an improper fraction (where the numerator is greater than or equal to the denominator) into a mixed number, divide the numerator by the denominator. The quotient becomes the whole number part, and the remainder over the original denominator forms the fractional part of the mixed number.

Formula To Remember

1. *Adding mixed numbers* ☞ $a\frac{b}{c} + d\frac{e}{f} = (a+d) + (\frac{b}{c} + \frac{e}{f}) = (a+d) + (\frac{bf+ec}{cf})$.
2. *Conversion from Improper Fraction to Mixed Number* ☞ $\frac{a}{b} = q\frac{r}{b}$ where q is the quotient and r is the remainder when a is divided by b.

1.5 Subtracting Mixed Numbers

Key Point

A mixed number $a\frac{c}{b}$ can be converted into an improper fraction by the formula:

$$a\frac{c}{b} = a + \frac{c}{b} = \frac{ab+c}{b}.$$

Key Point

To subtract mixed numbers, first convert them into improper fractions. Ensure the fractions have a common denominator. Subtract the fractions, simplify the result, and if needed, convert it back into a mixed number.

Formula To Remember

1. Convert to Improper Fractions $a\frac{c}{b} = \frac{ab+c}{b}$

1.6 Multiplying and Dividing Mixed Numbers

Key Point

Multiplication or division of mixed numbers involves first converting them into improper fractions, then multiplying or dividing, simplifying the result, and converting back to a mixed number if necessary.

2. Decimals

2.1 Comparing Decimals

Key Point

To compare two decimal numbers, we match each digit of the two decimals in the same place value, starting from the left. We compare the hundreds, tens, ones, tenth, hundredth, and so on.

2.2 Rounding Decimals

Key Point

To round a decimal to a certain place, find the digit in the next smallest place value (to the right of the place you are rounding to). If this digit is 5 or greater, add 1 to the digit in the place you are rounding to. If it is less than 5, the digit in the place you are rounding to remains unchanged.

2.3 Adding and Subtracting Decimals

Key Point

When adding or subtracting decimals, align the decimal points and ensure each number has the same number of decimal digits. Perform column addition or subtraction starting from the rightmost digit.

2.4 Multiplying and Dividing Decimals

> **Key Point**
>
> In multiplying decimals, ignore the decimal points initially, multiply as integers, then place the decimal in the product based on the total decimal places from the factors. For dividing decimals, make the divisor a whole number by adjusting the decimal points, and then divide as with whole numbers.

3. Integers and Order of Operations

3.1 Adding and Subtracting Integers

🔔 Key Point

To add integers, consider their signs. If both integers are positive or both are negative, add their values and keep the sign of the integers. If one is positive and the other negative, ignore the signs and subtract the smaller number from the larger number; the result takes the sign of the larger original number.

🔔 Key Point

To subtract an integer, add its opposite. For example, subtracting b from a is the same as adding the opposite of b to a:

$$a - b = a + (-b).$$

📌 Formula To Remember

1. Adding Two Negative Integers ☞ $(-a) + (-b) = -(a+b)$ (both a and b are positive).
2. Adding Positive and Negative Integers ☞ $a + (-b)$ is equal to $a - b$ if $a > b$, and is equal to $-(b - a)$ if $b > a$, where a and b are positive.
3. Subtracting Integers Using Addition ☞ $a - b = a + (-b)$

3.2 Multiplying and Dividing Integers

> **Key Point**
>
> Remember these rules for multiplying and dividing integers:
> - (negative) × (negative) = positive
> - (negative) ÷ (negative) = positive
> - (negative) × (positive) = negative
> - (negative) ÷ (positive) = negative
> - (positive) × (positive) = positive
> - (positive) ÷ (positive) = positive

3.3 Order of Operations

> **Key Point**
>
> The operations in mathematics must be performed in a specific order known as PEMDAS:
> - Parentheses
> - Exponents
> - Multiplication and Division (from left to right)
> - Addition and Subtraction (from left to right)
>
> You can remember PEMDAS using the phrase "Please Excuse My Dear Aunt Sally".

3.4 Absolute Value

> **Key Point**
>
> The absolute value of a number represents its distance from zero on the number line, regardless of direction. It is always non-negative. To find the absolute value of any number, consider only its magnitude, not its sign.

3.4 Absolute Value

Formula To Remember

1. *Definition of Absolute Value* 👉 $|x| = \begin{cases} x & \text{if } x \geq 0, \\ -x & \text{if } x < 0. \end{cases}$

2. *Multiplication involving Absolute Values* 👉 $|a \times b| = |a| \times |b|$.

3. *Division involving Absolute Values* 👉 $\left|\frac{a}{b}\right| = \frac{|a|}{|b|}$ where $b \neq 0$.

4. Ratios and Proportions

4.1 Simplifying Ratios

Key Point

A ratio represents the relative size of two or more quantities. The key to simplifying ratios is to divide all numbers by their greatest common divisor (GCD).

Formula To Remember

1. *Simplification by GCD* 👉 The simplified form of the ratio $a:b$ is $\frac{a \div \gcd(a,b)}{b \div \gcd(a,b)} = \frac{a'}{b'}$ where $\gcd(a,b)$ is the Greatest Common Divisor of a and b.

4.2 Proportional Ratios

Key Point

Cross-multiplication simplifies solving proportions. You multiply diagonally across the equal sign to find missing numbers.

Formula To Remember

1. *Proportion Equation* 👉 $\frac{a}{b} = \frac{c}{d}$ is the same as $a:b = c:d$.
2. *Cross-Multiplication Method* 👉 $a:b = c:d$ is equivalent to $a \times d = b \times c$.

4.3 Similarity and Ratios

> **Key Point**
>
> Two figures are similar if they have the same shape. This means that all their corresponding angles are equal and corresponding sides are in proportion.

5. Percentage

5.1 Percent Problems

Key Point

Three key entities in any percent problem are the "percent", the "base", and the "part".

- Percent: The ratio or the fraction of the quantity in comparison to 100.
- Base: The total number or the original quantity being considered.
- Part: The proportion or the fraction of the base.

Formula To Remember

1. *Finding the Part* 👉 Part = Percent × Base
2. *Finding the Percent* 👉 Percent = $\frac{Part}{Base} \times 100\%$
3. *Finding the Base* 👉 Base = $\frac{Part}{Percent}$

5.2 Percent of Increase and Decrease

Key Point

The formula for percent of change is:

$$\text{Percent of change} = \frac{\text{new number} - \text{original number}}{\text{original number}} \times 100.$$

If your answer is a negative number, then this signifies a percentage decrease. If it is positive, it is a percentage increase.

Formula To Remember

1. *Percent of Change* ☞ Percent of change = $\frac{\text{new number} - \text{original number}}{\text{original number}} \times 100$.

5.3 Discount, Tax, and Tip

Key Point

To calculate the discount and the selling price, use the following formulas:
- Discount = Original Price × Discount Rate,
- Selling Price = Original Price − Discount..

Key Point

To calculate the tax and the tip, use these formulas:
- Tax = Price × Tax Rate,
- Tip = Total Bill × Tip Rate.

Formula To Remember

1. *Calculating Discount* ☞ Discount = Original Price × Discount Rate
2. *Calculating Selling Price* ☞ Selling Price = Original Price − Discount
3. *Calculating Tax* ☞ Tax = Price × Tax Rate
4. *Calculating Tip* ☞ Tip = Total Bill × Tip Rate

5.4 Simple Interest

🔔 Key Point

Simple Interest is given by $I = prt$, where I is the interest, p the principal (initial amount), r the interest rate, and t the time (in years).

📌 Formula To Remember

1 *Simple Interest* 👉 $I = prt$ where I is the interest, p is the principal amount, r is the interest rate per year in decimal form, and t is the time in years.

6. Exponents and Variables

6.1 Multiplication Property of Exponents

Key Point

Exponents are shorthand for repeated multiplication of the same number.

Key Point

The rules for multiplying and simplifying exponents are:
1. When multiplying like bases, add the exponents: $x^a \times x^b = x^{a+b}$.
2. For a product of numbers raised to the same power: $x^a \times y^a = (xy)^a$.
3. When raising a power to another power, multiply the exponents: $(x^a)^b = x^{a \times b}$.

Formula To Remember

1. *Multiplying Like Bases* 👉 $x^a \times x^b = x^{a+b}$
2. *Product to a Power* 👉 $(xy)^a = x^a \times y^a$
3. *Power of a Power* 👉 $(x^a)^b = x^{a \times b}$

6.2 Division Property of Exponents

🔔 Key Point

The rules for dividing and simplifying exponents are:

1. When dividing like bases, subtract the exponents: $\frac{x^a}{x^b} = x^{a-b}$, with $x \neq 0$.
2. For a quotient of numbers raised to the same power: $\frac{x^a}{y^a} = \left(\frac{x}{y}\right)^a$, with $y \neq 0$.
3. If the exponent of the denominator is larger, rewrite the expression: $\frac{x^a}{x^b} = \frac{1}{x^{b-a}}$, with $x \neq 0$.

📌 Formula To Remember

1. *Dividing Like Bases* ☞ $\frac{x^a}{x^b} = x^{a-b}$ where $x \neq 0$.
2. *Quotient with Same Power* ☞ $\left(\frac{x}{y}\right)^a = \frac{x^a}{y^a}$ where $y \neq 0$.
3. *Larger Exponent in Denominator* ☞ $\frac{x^a}{x^b} = \frac{1}{x^{b-a}}$ where $x \neq 0$ and $b > a$.

6.3 Zero and Negative Exponents

🔔 Key Point

The Zero-Exponent Rule states that any non-zero number raised to the power of zero equals 1, $a^0 = 1$. However, 0^0 remains indeterminate.

🔔 Key Point

A negative exponent tells us to flip the fraction (take the reciprocal) and then use the positive exponent. So, if you have a fraction raised to a negative exponent, like $\left(\frac{a}{b}\right)^{-n}$, you flip the fraction to $\frac{b}{a}$ and then raise it to the positive exponent n:

$$\left(\frac{a}{b}\right)^{-n} = \left(\frac{b}{a}\right)^n.$$

Formula To Remember

1. *Zero Exponent Rule* — $a^0 = 1$ for any non-zero number a.
2. *Negative Exponent Rule* — $a^{-n} = \frac{1}{a^n}$ for any non-zero number a and positive integer n.
3. *Reciprocal Property for Fractions* — $\left(\frac{a}{b}\right)^{-n} = \left(\frac{b}{a}\right)^n$ for any non-zero numbers a and b, and positive integer n.

6.4 Working with Negative Bases

Key Point

For $(-a)^n$, the expression equals a^n when n is an even positive integer and $-a^n$ when n is an odd positive integer.

Formula To Remember

1. *Negative Base to Even Power* — $(-a)^{2n} = a^{2n}$ where n is a positive integer.
2. *Negative Base to Odd Power* — $(-a)^{2n+1} = -a^{2n+1}$ where n is a non-negative integer.

6.5 Scientific Notation

Key Point

Scientific notation expresses numbers as $m \times 10^n$, where $1 \leq m < 10$ is the mantissa and n is an integer exponent.

Key Point

To convert from scientific to standard notation, shift the decimal point in m right for positive n and left for negative n.

6.6 Radicals

🔔 Key Point

The notation $\sqrt[n]{x}$ signifies the nth root of x, which is defined as the value y such that $y^n = x$. This relationship is algebraically expressed as $y = x^{\frac{1}{n}}$.

🔔 Key Point

To add or subtract radicals, the terms underneath the radical must be the same. Hence, similar to like terms, only like radicals can be added or subtracted.

🔔 Key Point

Rules for operations with radicals:
- Multiply radicals with the same root: $\sqrt[n]{x} \times \sqrt[n]{y} = \sqrt[n]{xy}$.
- Divide radicals with the same root: $\frac{\sqrt[n]{x}}{\sqrt[n]{y}} = \sqrt[n]{\frac{x}{y}}$, for $y \neq 0$.
- Raise a radical to a power: $(\sqrt[n]{x})^m = \sqrt[n]{x^m}$.

📌 Formula To Remember

1. Definition of a Radical ☞ $\sqrt[n]{x} = x^{\frac{1}{n}}$
2. Adding/Subtracting Like Radicals ☞ $a\sqrt[n]{x} \pm b\sqrt[n]{x} = (a \pm b)\sqrt[n]{x}$
3. Multiplication of Radicals (Same Index) ☞ $\sqrt[n]{x} \times \sqrt[n]{y} = \sqrt[n]{xy}$
4. Division of Radicals (Same Index) ☞ $\frac{\sqrt[n]{x}}{\sqrt[n]{y}} = \sqrt[n]{\frac{x}{y}}$ for $y \neq 0$
5. Raising a Radical to a Power ☞ $(\sqrt[n]{x})^m = \sqrt[n]{x^m}$

7. Expressions and Variables

7.1 Simplifying Algebraic Expressions

Key Point

In algebra, terms with identical variables and exponents, known as "like" terms, can be combined to simplify expressions.

Formula To Remember

1. *Combine Like Terms* 👉 $ax^n + bx^n = (a+b)x^n$ where a and b are coefficients and n is the power of x.

2. *Simplifying with Fractional Exponents* 👉 $ax^{\frac{m}{n}} + bx^{\frac{m}{n}} = (a+b)x^{\frac{m}{n}}$ where a and b are coefficients, m and n are integers.

7.2 Simplifying Polynomial Expressions

Key Point

To simplify a polynomial, combine 'like' terms. Ensure that the simplified expression is presented in standard form, with terms arranged in descending order of their degree (the highest power of the variable in the term).

Formula To Remember

1 *Standard Form of a Polynomial* 👉 Arrange the expression in descending order of power: $a_n x^n + a_{n-1} x^{n-1} + \ldots + a_0$.

7.3 The Distributive Property

🔔 Key Point

The rule of the Distributive Property is represented as:

$$a(b+c) = ab + ac, \quad a(b-c) = ab - ac.$$

Formula To Remember

1 *Distributive Property over Addition* 👉 $a(b+c) = ab + ac$

2 *Distributive Property over Subtraction* 👉 $a(b-c) = ab - ac$

7.4 Evaluating One Variable

🔔 Key Point

To evaluate an expression with a variable, first substitute the given value for the variable, then apply the order of operations: parentheses, exponents, multiplication and division (left to right), and addition and subtraction (left to right).

7.5 Evaluating Two Variables

🔔 Key Point

Always remember to follow the order of operations while evaluating the expression after substitution: Parentheses, Exponents, Multiplication and Division (from left to right), Addition and Subtraction (from left to right).

8. Equations and Inequalities

8.1 One-Step Equations

🔔 Key Point

Solving one-step equations refers to finding the value of the unknown variable by performing a single mathematical operation.

🔔 Key Point

The inverse operation is the reverse of the operation being performed. Identifying and understanding the inverse operation is crucial for solving one-step equations.

📌 Formula To Remember

1. *Solving Addition Equations* 👉 $x + a = b \Rightarrow x = b - a$
2. *Solving Subtraction Equations* 👉 $x - a = b \Rightarrow x = b + a$
3. *Solving Multiplication Equations* 👉 $ax = b \Rightarrow x = \frac{b}{a}$
4. *Solving Division Equations* 👉 $\frac{x}{a} = b \Rightarrow x = ab$

8.2 Multi-Step Equations

🔔 Key Point

Solving multi-step equations involves combining like terms, rearranging terms to isolate the variable, using inverse operations to simplify, and always verifying the solution by substituting it back into the original equation.

📌 Formula To Remember

1. Combine Like Terms 👉 $ax + bx = (a+b)x$
2. Variables on One Side 👉 $ax \pm c = bx \pm d \Rightarrow ax - bx = \pm(d-c)$
3. Isolate Variable using Addition/Subtraction 👉 $ax \pm b = c \Rightarrow ax = c \mp b$
4. Isolate Variable using Multiplication/Division 👉 $ax = b \Rightarrow x = \frac{b}{a}$

8.3 System of Equations

🔔 Key Point

A system of equations consists of multiple equations with the same variables.

🔔 Key Point

The elimination method involves adding or subtracting the equations in the system in order to eliminate one of the variables, allowing for simpler computation.

8.4 One-Step Inequalities

🔔 Key Point

To solve one-step inequalities, apply the inverse operation to both sides.

🔔 Key Point

If you multiply or divide both sides of an inequality by a negative number, you must flip the direction of the inequality sign. This step is important to keep the inequality true.

8.5 Multi-Step Inequalities

Key Point

To solve multi-step inequalities, simplify each side by combining like terms and moving all variables to one side using addition or subtraction. Then, isolate the variable with inverse operations.

Key Point

Always reverse the inequality sign when you multiply or divide both sides by a negative number.

Formula To Remember

1. *Moving Variables to One Side* ☞ $ax \pm b \leq c \pm dx \Rightarrow (a \mp d)x \leq c \mp b$
2. *Isolate the Variable Using Multiplication/Division* ☞ $bx \leq d \Rightarrow x \leq \frac{d}{b},\ b > 0$
3. *Reverse Inequality When Multiplying/Dividing by a Negative* ☞ $bx \leq d \Rightarrow x \geq \frac{d}{b},\ b < 0$

8.6 Graphing Single-Variable Inequalities

Key Point

When graphing single-variable inequalities on a number line: use an open circle for "$<$" or "$>$" and a filled circle for "$\leq$" or "$\geq$." Point the arrow right for "greater than" inequalities and left for "less than" inequalities.

9. Lines and Slope

9.1 Finding Slope

Key Point

In a coordinate plane, a line is defined by its points and characterized by its steepness and direction. The plane consists of two perpendicular axes (x and y) that intersect at the origin $(0,0)$.

Key Point

The slope of a line between two points $A(x_1, y_1)$ and $B(x_2, y_2)$ is found using the formula:

$$\text{slope} = \frac{y_2 - y_1}{x_2 - x_1},$$

This represents the ratio of the vertical change (*rise*) to the horizontal change (*run*).

Formula To Remember

1 *Slope from Two Points* 👉 Given two points $A(x_1, y_1)$ and $B(x_2, y_2)$, the slope of the line through A and B is: $\text{slope} = \frac{y_2 - y_1}{x_2 - x_1}$

9.2 Graphing Lines Using Slope-Intercept Form

Key Point

The slope-intercept form of a line is $y = mx + b$, where m is the slope and b is the y-intercept.

9.3 Writing Linear Equations

 Formula To Remember

1. *Slope-Intercept Form of a Line* 👉 $y = mx + b$ where m is the slope and b is the y-intercept.
2. *Finding the y-Intercept* 👉 Set $x = 0$ in the equation $y = mx + b$ to find the y-intercept $(0, b)$.

9.3 Writing Linear Equations

🔔 **Key Point**

To write a linear equation, start by identifying the slope (m) and then find the y-intercept (b).

These are the key components of the linear equation format: $y = mx + b$.

9.4 Finding Midpoint

🔔 **Key Point**

The formula for the midpoint, M, of two endpoints $A(x_1, y_1)$ and $B(x_2, y_2)$ is as follows:

$$M = \left(\frac{x_1 + x_2}{2}, \frac{y_1 + y_2}{2}\right).$$

 Formula To Remember

1. *Midpoint Formula* 👉 $M = \left(\frac{x_1+x_2}{2}, \frac{y_1+y_2}{2}\right)$ for endpoints $A(x_1, y_1)$ and $B(x_2, y_2)$.

9.5 Finding Distance of Two Points

🔔 **Key Point**

The formula to find the distance d between two points (x_1, y_1) and (x_2, y_2) is:

$$d = \sqrt{(x_2 - x_1)^2 + (y_2 - y_1)^2}.$$

 Formula To Remember

1. *Distance Formula* 👉 $d = \sqrt{(x_2 - x_1)^2 + (y_2 - y_1)^2}$

9.6 Graphing Linear Inequalities

Key Point

When graphing linear inequalities, use a dashed line for "$<$" or "$>$" to exclude the boundary line, and a solid line for "$\leq$" or "$\geq$" to include the boundary line in the solution region.

10. Polynomials

10.1 Simplifying Polynomials

Key Point

A polynomial $P(x)$ is an expression of the form $a_n x^n + a_{n-1} x^{n-1} + \ldots + a_1 x + a_0$, where $a_n, a_{n-1}, \ldots, a_0$ are coefficients, and n is the degree, indicating the highest power of x.

Key Point

"Like" terms in a polynomial are those which have the same variables raised to the same power.

Formula To Remember

1. General Form of a Polynomial $P(x) = a_n x^n + a_{n-1} x^{n-1} + \ldots + a_1 x + a_0$

10.2 Adding and Subtracting Polynomials

Key Point

To add or subtract polynomials, combine the 'like' terms. Like terms are terms that have the same variable and the same power. When adding, you sum their coefficients, and when subtracting, you subtract the coefficients of the like terms.

Formula To Remember

1. Addition of Polynomials $(a_n x^n + a_{n-1} x^{n-1} + \cdots + a_1 x + a_0) + (b_n x^n + b_{n-1} x^{n-1} + \cdots + b_1 x + b_0) = (a_n + b_n) x^n + (a_{n-1} + b_{n-1}) x^{n-1} + \cdots + (a_1 + b_1) x + (a_0 + b_0)$

2. Subtraction of Polynomials $(a_n x^n + a_{n-1} x^{n-1} + \cdots + a_1 x + a_0) - (b_n x^n + b_{n-1} x^{n-1} + \cdots + b_1 x + b_0) = (a_n - b_n) x^n + (a_{n-1} - b_{n-1}) x^{n-1} + \cdots + (a_1 - b_1) x + (a_0 - b_0)$

10.3 Multiplying and Dividing Monomials

Key Point

A monomial is a polynomial with only one term. Examples include $3x^2$ and $-4x^2 y$.

Formula To Remember

1. Multiplication of Monomials $c_1 x^a y^b \times c_2 x^m y^n = (c_1 \times c_2) x^{a+m} y^{b+n}$

2. Division of Monomials $\dfrac{c_1 x^a y^b}{c_2 x^m y^n} = \left(\dfrac{c_1}{c_2}\right) x^{a-m} y^{b-n}$ where $c_2 \neq 0$

10.4 Multiplying a Polynomial and a Monomial

Key Point

The distributive property: When you multiply a polynomial with a monomial, you multiply the monomial by each term in the polynomial individually, rather than to the polynomial as a whole.

10.5 Multiplying Binomials

Key Point

To multiply two binomials together, we employ the "FOIL" method, an acronym for First, Outer, Inner, Last, a stepwise process that multiplies each term together with every other term.

Formula To Remember

1. The FOIL Method $(x+a)(x+b) = x^2 + bx + ax + ab = x^2 + (a+b)x + ab$.

10.6 Factoring Trinomials

🔔 Key Point

The "Reverse FOIL" method is used to factor a trinomial into two binomials. For a trinomial like $x^2 + (b+a)x + ab$, you need to find two numbers, a and b, that multiply to ab and add up to $(b+a)$. These numbers are then used to form the binomials $(x+a)$ and $(x+b)$.

🔔 Key Point

The difference of two squares, $a^2 - b^2$, can be factored as the product of $(a+b)$ and $(a-b)$.

📌 Formula To Remember

1. *Reverse FOIL Method* 👉 $x^2 + (a+b)x + ab = (x+a)(x+b)$
2. *Difference of Squares* 👉 $a^2 - b^2 = (a+b)(a-b)$
3. *Perfect Square Trinomial (Sum)* 👉 $a^2 + 2ab + b^2 = (a+b)^2$
4. *Perfect Square Trinomial (Difference)* 👉 $a^2 - 2ab + b^2 = (a-b)^2$

11. Geometry and Solid Figures

11.1 Complementary and Supplementary angles

Key Point

Two angles are said to be complementary if their measures add up to 90°. In expression form, we write: if x and y are complementary, then $x+y=90°$.

Key Point

Two angles are said to be supplementary if their measures add up to 180°. In expression form: if x and y are supplementary, then $x+y=180°$.

Formula To Remember

1. *Complementary Angle Formula* 👉 If x and y are complementary, then $x+y=90°$.
2. *Supplementary Angle Formula* 👉 If x and y are supplementary, then $x+y=180°$.

11.2 Parallel Lines and Transversals

Key Point

When a transversal intersects two parallel lines, it forms two sets of congruent angles.

Key Point

When a transversal intersects parallel lines, it also forms sets of supplementary angles.

11.3 Triangles

> **Key Point**
>
> In any triangle, the sum of all angles equals 180°.

> **Key Point**
>
> The area of a triangle is $\frac{1}{2} \times (base \times height)$.

> **Formula To Remember**
>
> 1. *Angle Sum Property* 👉 In any triangle, the sum of the interior angles is always 180°: $A + B + C = 180°$.
> 2. *Area of a Triangle* 👉 The area A of a triangle with base b and height h is given by: $A = \frac{1}{2} \times b \times h$.

11.4 The Pythagorean Theorem

> **Key Point**
>
> The Pythagorean Theorem states $a^2 + b^2 = c^2$, where c is the hypotenuse, and a and b are the lengths of the other two sides of a right-angled triangle.

> **Formula To Remember**
>
> 1. *Pythagorean Theorem* 👉 In a right-angled triangle, the square of the length of the hypotenuse (c) is equal to the sum of the squares of the other two sides (a and b): $a^2 + b^2 = c^2$

11.5 Special Right Triangles

> **Key Point**
>
> The sides of a $45° - 45° - 90°$ triangle are in the ratio $1 : 1 : \sqrt{2}$. This means that the legs of the triangle are congruent, and the hypotenuse is $\sqrt{2}$ times as long as one leg.

Key Point

The sides of a 30° − 60° − 90° triangle are in the ratio $1 : \sqrt{3} : 2$. The side opposite the 30° angle is the shortest, the side opposite the 60° angle is $\sqrt{3}$ times longer, and the hypotenuse is twice as long as the shortest side.

Formula To Remember

1. 45° − 45° − 90° *Triangle Side Ratios* ☞ For 45° − 45° − 90° triangle: sides are $a, a, a\sqrt{2}$
2. 30° − 60° − 90° *Triangle Side Ratios* ☞ For 30° − 60° − 90° triangle: sides are $a, a\sqrt{3}, 2a$

11.6 Polygons

Key Point

The perimeter of a square is four times the length of one side, $P_{\text{square}} = 4s$. The area is the side length squared, $A_{\text{square}} = s^2$.

Key Point

The perimeter of a rectangle is the sum of twice its width and length, $P_{\text{rectangle}} = 2(\text{width} + \text{length})$. The area is the product of its width and length, $A_{\text{rectangle}} = \text{width} \times \text{length}$.

Key Point

The perimeter of a trapezoid is the sum of all its side lengths. The area is half the product of the height and the sum of the parallel sides, $A_{\text{trapezoid}} = \frac{1}{2} \times (\text{base}_1 + \text{base}_2) \times \text{height}$.

Key Point

The perimeter of a hexagon is six times a single side length, $P_{\text{hexagon}} = 6s$. The area is calculated using the formula $A_{\text{hexagon}} = \frac{3\sqrt{3}}{2} \times s^2$.

Key Point

The perimeter of a parallelogram is twice the sum of adjacent side lengths. Its area is the product of the base and the height perpendicular to it, $A_{\text{parallelogram}} = \text{base} \times \text{height}$.

Formula To Remember

1. Perimeter of a Square ☞ $P_{square} = 4s$
2. Area of a Square ☞ $A_{square} = s^2$
3. Perimeter of a Rectangle ☞ $P_{rectangle} = 2(\text{width} + \text{length})$
4. Area of a Rectangle ☞ $A_{rectangle} = \text{width} \times \text{length}$
5. Perimeter of a Trapezoid ☞ $P_{trapezoid} = base_1 + base_2 + leg_1 + leg_2$
6. Area of a Trapezoid ☞ $A_{trapezoid} = \frac{1}{2} \times (base_1 + base_2) \times \text{height}$
7. Perimeter of a Hexagon ☞ $P_{hexagon} = 6s$
8. Area of a Regular Hexagon ☞ $A_{hexagon} = \frac{3\sqrt{3}}{2} \times s^2$
9. Perimeter of a Parallelogram ☞ $P_{parallelogram} = 2(a+b)$
10. Area of a Parallelogram ☞ $A_{parallelogram} = \text{base} \times \text{height}$

11.7 Circles

Key Point

A circle's diameter is twice its radius, and its circumference is 2π times its radius.

Key Point

The equation for the area of a circle is $A = \pi r^2$, and the equation for the circumference of a circle is $C = 2\pi r$.

Formula To Remember

1. Diameter and Radius ☞ $d = 2r$ where d is the diameter, and r is the radius of the circle.
2. Circumference of a Circle ☞ $C = 2\pi r$ where C is the circumference, r is the radius, and $\pi \approx 3.14$.
3. Area of a Circle ☞ $A = \pi r^2$ where A is the area, r is the radius, and $\pi \approx 3.14$.

11.8 Cubes

Key Point

Volume is the measure of the amount of space inside of a solid figure.

Key Point

For a cube with side length a, the volume is $V = a^3$ and the surface area is $S = 6a^2$.

Formula To Remember

1. *Volume of a Cube* 👉 $V = a^3$ where a is the length of a side of the cube.
2. *Surface Area of a Cube* 👉 $A_s = 6a^2$ where a is the length of a side of the cube.

11.9 Rectangular Prisms

Key Point

A rectangular prism is a three-dimensional shape which consists of six rectangular faces.

Key Point

The volume of a rectangular prism is calculated by multiplying its length, width, and height. This formula provides the measure of space the prism occupies.

Key Point

The surface area of a Rectangular prism equals twice the sum of the areas of the length-width, width-height, and length-height faces.

Formula To Remember

1. *Volume of a Rectangular Prism* 👉 $V = l \times w \times h$
2. *Surface Area of a Rectangular Prism* 👉 $A = 2 \times (wh + lw + lh)$

11.10 Cylinder

Key Point

A cylinder is a solid figure with straight parallel sides and a circular or oval cross-section.

11.10 Cylinder

🔔 Key Point

The volume of a cylinder is calculated using $\pi r^2 h$, and its surface area is given by $2\pi r^2 + 2\pi rh$, where r is the radius and h is the height of the cylinder.

📌 Formula To Remember

1. *Volume of a Cylinder* 👉 $V = \pi r^2 h$ where V is the volume, r is the radius, and h is the height of the cylinder.

2. *Surface Area of a Cylinder* 👉 $A = 2\pi r(r+h)$ where A is the surface area, r is the radius, and h is the height of the cylinder.

12. Statistics

12.1 Mean, Median, Mode, and Range of the Given Data

> **Key Point**
>
> When all data entries are equally important, the mean is a good representation of the central tendency of the data. It is calculated as Mean $= \frac{\sum_{i=1}^{n} x_i}{n}$, where $\sum_{i=1}^{n} x_i$ is the sum of all data values and n is the number of data entries.

> **Key Point**
>
> The mode provides information about the most frequent observation in a data set. It is the value that appears most frequently.

> **Key Point**
>
> The median provides a measure of central location that is not affected by outliers or extreme values. It is the middle value when the data set is ordered. If there is an even number of observations, the median is the average of the two middle numbers: Median $= \frac{x_{\frac{n}{2}} + x_{\frac{n}{2}+1}}{2}$ for an even number of observations n.

12.2 Pie Graph

> **Key Point**
>
> The range provides a measure of dispersion or variation in a data set. It is calculated as Range = Max(X) − Min(X), representing the difference between the maximum and minimum values in the data set X.

> **Formula To Remember**
>
> 1. *Calculating Mean* ☞ Mean $= \frac{x_1+x_2+\cdots+x_n}{n}$
> 2. *Determining Mode* ☞ Mode is the value(s) that occurs most frequently in the data set.
> 3. *Determining Median (Odd count of numbers)* ☞ Median $= x_{\frac{n+1}{2}}$
> 4. *Determining Median (Even count of numbers)* ☞ Median $= \frac{x_{\frac{n}{2}}+x_{\frac{n}{2}+1}}{2}$
> 5. *Calculating Range* ☞ Range = Largest data value − Smallest data value

12.2 Pie Graph

> **Key Point**
>
> Each sector in a pie chart corresponds to a category of data. The larger the sector, the greater the relative proportion that category represents in the total data set.

> **Formula To Remember**
>
> 1. *Calculate Percentage of Category* ☞ Percentage $= \left(\frac{\text{Number of instances in category}}{\text{Total number of instances}}\right) \times 100\%$
> 2. *Calculate Angle of Sector Using Degrees* ☞ Angle in degrees $= \left(\frac{\text{Number of instances in category}}{\text{Total number of instances}}\right) \times 360°$
> 3. *Calculate Quantity from Percentage* ☞ Quantity = Percentage × Total quantity

12.3 Probability Problems

> **Key Point**
>
> Probability is always expressed as a number between zero (the event will definitely not happen) and 1 (the event will definitely happen).

Key Point

The Probability of an event E is given by the formula:

$$P(E) = \frac{\text{number of desired outcomes}}{\text{number of total outcomes}}.$$

Formula To Remember

1. Basic Probability Formula 👉 $P(E) = \frac{\text{number of desired outcomes}}{\text{number of total outcomes}}$

12.4 Permutations and Combinations

Key Point

Factorials are the product of an integer and all the positive integers below it. It is denoted as $n!$, for example, $4! = 4 \times 3 \times 2 \times 1$.

Key Point

Permutations (denoted as $P(n,r)$) count the different arrangements of a subset of r items from a set of n items, where the order matters. The formula is $P(n,r) = \frac{n!}{(n-r)!}$.

Key Point

Combinations (denoted as $C(n,r)$) count the different groups that can be formed from a set of n items, taking r at a time, where the order doesn't matter. The formula is $C(n,r) = \frac{n!}{r!(n-r)!}$.

Key Point

When repetition is allowed, the number of possible combinations can be calculated using the formula n^r, where n is the number of choices for each position, and r is the number of positions.

12.4 Permutations and Combinations

Formula To Remember

1. *Factorial Definition* 👉 $n! = n \times (n-1) \times (n-2) \times \cdots \times 2 \times 1$ for $n \geq 1$, with $0! = 1$.

2. *Permutations Formula* 👉 $P(n,r) = \frac{n!}{(n-r)!}$ for selecting r different items out of n where order matters.

3. *Combinations Formula* 👉 $C(n,r) = \frac{n!}{r!(n-r)!}$ for selecting r different items out of n where order does not matter.

4. *Permutations with Repetition* 👉 n^r where n is the number of choices for each position and r is the total number of positions.

13. Functions Operations

13.1 Function Notation and Evaluation

🔔 Key Point

Functions are mathematical operations that assign unique outputs to given inputs.

🔔 Key Point

When we evaluate functions, we substitute specific values of x into the function formula to find the corresponding output, $f(x)$.

📌 Formula To Remember

1. *Function Notation* 👉 $f(x)$ represents the function value at x.
2. *Evaluating a Function* 👉 To evaluate $f(x)$ at $x = a$, replace x with a: $f(a) = f(x)|_{x=a}$.

13.2 Adding and Subtracting Functions

🔔 Key Point

Adding or subtracting functions involves combining the corresponding terms of the functions. The result of this operation is a new function.

Formula To Remember
1. *Adding Functions* — $(f+g)(x) = f(x) + g(x)$
2. *Subtracting Functions* — $(f-g)(x) = f(x) - g(x)$

13.3 Multiplying and Dividing Functions

Key Point

When multiplying functions, you combine them to form $(f \cdot g)(x) = f(x) \cdot g(x)$, and for division, $\left(\frac{f}{g}\right)(x) = \frac{f(x)}{g(x)}$, ensuring $g(x) \neq 0$. These operations yield new functions, each defined based on the original functions.

Formula To Remember
1. *Multiplication of Functions* — $(f \cdot g)(x) = f(x) \cdot g(x)$
2. *Division of Functions* — $\left(\frac{f}{g}\right)(x) = \frac{f(x)}{g(x)}$, provided $g(x) \neq 0$

13.4 Composition of Functions

Key Point

The notation used for composition is:

$$(f \circ g)(x) = f(g(x)),$$

which is read as "f composed with g of x" or "f of g of x".

Formula To Remember
1. *Composition of Functions* — $(f \circ g)(x) = f(g(x))$

14. ASVAB Test Review and Strategies

14.1 The ASVAB Test Review and Scoring

Introduced in 1968, the *Armed Services Vocational Aptitude Battery (ASVAB)* has been taken by over 40 million individuals to date. As detailed on the official ASVAB website, this test serves as a comprehensive multi-aptitude assessment, gauging developed skills and forecasting potential academic and occupational achievements within the military. Annually, it's administered to over a million candidates, including military applicants, high school, and post-secondary students.

There are three formats of ASVAB:

- **The CAT-ASVAB:** A computer adaptive test, adjusts the difficulty level of questions based on the test-taker's responses. A correct answer leads to a more challenging subsequent question, whereas an incorrect response results in an easier one. Once an answer is selected, it cannot be altered.
- **The MET-site ASVAB:** A traditional paper and pencil test, administered at Military Entrance Testing (MET) sites.
- **The Student ASVAB:** Also a paper and pencil format, designed specifically for high school and post-secondary students.

Each version of the ASVAB is tailored to suit different testing environments and requirements, ensuring a broad assessment of aptitudes relevant to various military and educational pathways.

ASVAB results are presented in percentile scores ranging from 1 to 99, indicating how a test-taker's score compares to that of a reference group. For instance, a score of 90 suggests that the examinee performed as well as or better than 90% of a nationally-representative sample of test-takers, while a score of 60 means the examinee scored as well as or better than 60% of this group.

14.2 ASVAB Math Test-Taking Strategies

Successfully navigating the ASVAB Math test requires not only a solid understanding of mathematical concepts but also effective problem-solving strategies. In this section, we explore a range of strategies to optimize your performance and outcomes on the ASVAB Math test. From comprehending the question and using informed guessing to finding ballpark answers and employing backsolving and numeric substitution, these strategies will empower you to tackle various types of math problems with confidence and efficiency.

#1 Understand the Questions and Review Answers

Below are a set of effective strategies to optimize your performance and outcomes on the ASVAB Math test.

- **Comprehend the Question:** Begin by carefully reviewing the question to identify keywords and essential information.
- **Mathematical Translation:** Translate the identified keywords into mathematical operations that will enable you to solve the problem effectively.
- **Analyze Answer Choices:** Examine the answer choices provided and identify any distinctions or patterns among them.
- **Visual Aids:** If necessary, consider drawing diagrams or labeling figures to aid in problem-solving.
- **Pattern Recognition:** Look for recurring patterns or relationships within the problem that can guide your solution.
- **Select the Right Method:** Determine the most suitable strategies for answering the question, whether it involves straightforward mathematical calculations, numerical substitution (plugging in numbers), or testing the answer choices (backsolving); see below for a comprehensive explanation of these methods.
- **Verification:** Before finalizing your answer, double-check your work to ensure accuracy and completeness.

Let's review some of the important strategies in detail.

#2 Use Educated Guessing

This strategy is particularly useful for tackling problems that you have some understanding of but cannot solve through straightforward mathematics. In such situations, aim to eliminate as many answer choices as possible before making a selection. When faced with a problem that seems entirely unfamiliar, there's no need to spend excessive time attempting to eliminate answer choices. Instead, opt for a random choice before proceeding to the next question.

As you can see, employing direct solutions is the most effective approach. Carefully read the question, apply the math concepts you've learned, and align your answer with one of the available choices. Feeling stuck? Make your best-educated guess and move forward.

Never leave questions unanswered! Even if a problem appears insurmountable, make an effort to provide a response. If necessary, make an educated guess. Remember, you won't lose points for an incorrect answer, but you may earn points for a correct one!

#3 Ballpark Estimates

A *"ballpark estimate"* is a *rough approximation*. When dealing with complex calculations and numbers, it's easy to make errors. Sometimes, a small decimal shift can turn a correct answer into an incorrect one, no matter how many steps you've taken to arrive at it. This is where ballparking can be incredibly useful.

If you have an idea of what the correct answer might be, even if it's just a rough estimate, you can often eliminate a few answer choices. While answer choices typically account for common student errors and closely related values, you can still rule out choices that are significantly off the mark. When facing a multiple-choice question, deliberately look for answers that don't even come close to the ballpark. This strategy effectively helps eliminate incorrect choices during problem-solving.

#4 Backsolving

A significant portion of questions on the ASVAB Math test are presented in multiple-choice format. Many test-takers find multiple-choice questions preferable since the correct answer is among the choices provided. Typically, you'll have four options to choose from, and your task is to determine the correct one. One effective approach for this is known as *"backsolving."*

14.2 ASVAB Math Test-Taking Strategies

As mentioned previously, solving questions directly is the most optimal method. Begin by thoroughly examining the problem, calculating a solution, and then matching the answer with one of the available choices. However, if you find yourself unable to calculate a solution, the next best approach involves employing "backsolving."

When employing backsolving, compare one of the answer choices to the problem at hand and determine which choice aligns most closely. Frequently, answer choices are arranASVAB in either ascending or descending order. In such cases, consider testing options B or C first. If neither is correct, you can proceed either up or down from there.

#5 Plugging In Numbers

Using numeric substitution or *'plugging in numbers'* is a valuable strategy applicable to a wide array of math problems encountered on the ASVAB Math test. This approach is particularly helpful in simplifying complex questions, making them more manageable and comprehensible. By employing this strategy thoughtfully, you can arrive at the solution with ease.

The concept is relatively straightforward. Simply replace unknown variables in a problem with specific values. When selecting a number for substitution, consider the following guidelines:

- Opt for a basic number (though not overly basic). It's generally advisable to avoid choosing 1 (or even 0). A reasonable choice often includes selecting the number 2.
- Avoid picking a number already present in the problem statement.
- Ensure that the chosen numbers are distinct when substituting at least two of them.
- Frequently, the use of numeric substitution helps you eliminate some of the answer choices, so it's essential not to hastily select the first option that appears to be correct.
- When faced with multiple seemingly correct answers, you may need to opt for a different set of values and reevaluate the choices that haven't been ruled out yet.
- If your problem includes fractions, a valid solution might require consideration of either *the least common denominator (LCD)* or a multiple of the LCD.
- When tackling problems related to percentages, it's advisable to select the number 100 for numeric substitution.

15. Formula Sheet for All Topics

For your benefit we are providing all formulas again in one place. This is a quick reference guide to the formulas you need to remember for the math test. We recommend that you review this formula sheet before taking the practice tests. This will help you recall and apply the formulas effectively.

Formulas for Chapter: Fractions and Mixed Numbers

Simplifying Fractions

1. *Simplification by Common Divisor* ☞ $\frac{a}{b} = \frac{a \div d}{b \div d} = \frac{a'}{b'}$, where $d > 1$ is a common divisor of a and b.
2. *Simplification using GCD* ☞ $\frac{a}{b} = \frac{a \div \gcd(a,b)}{b \div \gcd(a,b)} = \frac{a''}{b''}$, where $\gcd(a,b)$ is the greatest common divisor of a and b.

Adding and Subtracting Fractions

1. *Adding 'Like' Fractions* ☞ $\frac{a}{b} + \frac{c}{b} = \frac{a+c}{b}$
2. *Subtracting 'Like' Fractions* ☞ $\frac{a}{b} - \frac{c}{b} = \frac{a-c}{b}$
3. *Adding 'Unlike' Fractions* ☞ $\frac{a}{b} + \frac{c}{d} = \frac{ad+bc}{bd}$
4. *Subtracting 'Unlike' Fractions* ☞ $\frac{a}{b} - \frac{c}{d} = \frac{ad-bc}{bd}$

Multiplying and Dividing Fractions

1. *Multiplication of Fractions* 👉 $\frac{a}{b} \times \frac{c}{d} = \frac{a \times c}{b \times d}$.

2. *Division of Fractions (Keep, Change, Flip)* 👉 $\frac{a}{b} \div \frac{c}{d} = \frac{a}{b} \times \frac{d}{c} = \frac{a \times d}{b \times c}$.

Adding Mixed Numbers

1. *Adding mixed numbers* 👉 $a\frac{b}{c} + d\frac{e}{f} = (a+d) + (\frac{b}{c} + \frac{e}{f}) = (a+d) + (\frac{bf+ec}{cf})$.

2. *Conversion from Improper Fraction to Mixed Number* 👉 $\frac{a}{b} = q\frac{r}{b}$ where q is the quotient and r is the remainder when a is divided by b.

Subtracting Mixed Numbers

1. *Convert to Improper Fractions* 👉 $a\frac{c}{b} = \frac{ab+c}{b}$

📖 Formulas For Chapter — Integers and Order of Operations

Adding and Subtracting Integers

1. *Adding Two Negative Integers* 👉 $(-a) + (-b) = -(a+b)$ (both a and b are positive).

2. *Adding Positive and Negative Integers* 👉 $a + (-b)$ is equal to $a - b$ if $a > b$, and is equal to $-(b-a)$ if $b > a$, where a and b are positive.

3. *Subtracting Integers Using Addition* 👉 $a - b = a + (-b)$

Absolute Value

1. *Definition of Absolute Value* 👉 $|x| = \begin{cases} x & \text{if } x \geq 0, \\ -x & \text{if } x < 0. \end{cases}$

2. *Multiplication involving Absolute Values* 👉 $|a \times b| = |a| \times |b|$.

3. *Division involving Absolute Values* 👉 $\left|\frac{a}{b}\right| = \frac{|a|}{|b|}$ where $b \neq 0$.

Formulas for Chapter: Ratios and Proportions

Simplifying Ratios

1. *Simplification by GCD* ☞ The simplified form of the ratio $a:b$ is $\frac{a \div \gcd(a,b)}{b \div \gcd(a,b)} = \frac{a'}{b'}$ where $\gcd(a,b)$ is the Greatest Common Divisor of a and b.

Proportional Ratios

1. *Proportion Equation* ☞ $\frac{a}{b} = \frac{c}{d}$ is the same as $a:b = c:d$.
2. *Cross-Multiplication Method* ☞ $a:b = c:d$ is equivalent to $a \times d = b \times c$.

Formulas for Chapter: Percentage

Percent Problems

1. *Finding the Part* ☞ Part = Percent × Base
2. *Finding the Percent* ☞ Percent = $\frac{\text{Part}}{\text{Base}} \times 100\%$
3. *Finding the Base* ☞ Base = $\frac{\text{Part}}{\text{Percent}}$

Percent of Increase and Decrease

1. *Percent of Change* ☞ Percent of change = $\frac{\text{new number} - \text{original number}}{\text{original number}} \times 100$.

Discount, Tax, and Tip

1. *Calculating Discount* ☞ Discount = Original Price × Discount Rate
2. *Calculating Selling Price* ☞ Selling Price = Original Price − Discount
3. *Calculating Tax* ☞ Tax = Price × Tax Rate
4. *Calculating Tip* ☞ Tip = Total Bill × Tip Rate

Simple Interest

1. *Simple Interest* 👉 $I = prt$ where I is the interest, p is the principal amount, r is the interest rate per year in decimal form, and t is the time in years.

Formulas For Chapter: Exponents and Variables

Multiplication Property of Exponents

1. *Multiplying Like Bases* 👉 $x^a \times x^b = x^{a+b}$
2. *Product to a Power* 👉 $(xy)^a = x^a \times y^a$
3. *Power of a Power* 👉 $(x^a)^b = x^{a \times b}$

Division Property of Exponents

1. *Dividing Like Bases* 👉 $\frac{x^a}{x^b} = x^{a-b}$ where $x \neq 0$.
2. *Quotient with Same Power* 👉 $\left(\frac{x}{y}\right)^a = \frac{x^a}{y^a}$ where $y \neq 0$.
3. *Larger Exponent in Denominator* 👉 $\frac{x^a}{x^b} = \frac{1}{x^{b-a}}$ where $x \neq 0$ and $b > a$.

Zero and Negative Exponents

1. *Zero Exponent Rule* 👉 $a^0 = 1$ for any non-zero number a.
2. *Negative Exponent Rule* 👉 $a^{-n} = \frac{1}{a^n}$ for any non-zero number a and positive integer n.
3. *Reciprocal Property for Fractions* 👉 $\left(\frac{a}{b}\right)^{-n} = \left(\frac{b}{a}\right)^n$ for any non-zero numbers a and b, and positive integer n.

Working with Negative Bases

1. *Negative Base to Even Power* 👉 $(-a)^{2n} = a^{2n}$ where n is a positive integer.
2. *Negative Base to Odd Power* 👉 $(-a)^{2n+1} = -a^{2n+1}$ where n is a non-negative integer.

Radicals

1. *Definition of a Radical* 👉 $\sqrt[n]{x} = x^{\frac{1}{n}}$
2. *Adding/Subtracting Like Radicals* 👉 $a\sqrt[n]{x} \pm b\sqrt[n]{x} = (a \pm b)\sqrt[n]{x}$
3. *Multiplication of Radicals (Same Index)* 👉 $\sqrt[n]{x} \times \sqrt[n]{y} = \sqrt[n]{xy}$
4. *Division of Radicals (Same Index)* 👉 $\frac{\sqrt[n]{x}}{\sqrt[n]{y}} = \sqrt[n]{\frac{x}{y}}$ for $y \neq 0$
5. *Raising a Radical to a Power* 👉 $(\sqrt[n]{x})^m = \sqrt[n]{x^m}$

Formulas for Chapter: Expressions and Variables

Simplifying Algebraic Expressions

1. *Combine Like Terms* 👉 $ax^n + bx^n = (a+b)x^n$ where a and b are coefficients and n is the power of x.
2. *Simplifying with Fractional Exponents* 👉 $ax^{\frac{m}{n}} + bx^{\frac{m}{n}} = (a+b)x^{\frac{m}{n}}$ where a and b are coefficients, m and n are integers.

Simplifying Polynomial Expressions

1. *Standard Form of a Polynomial* 👉 Arrange the expression in descending order of power: $a_n x^n + a_{n-1} x^{n-1} + \ldots + a_0$.

The Distributive Property

1. *Distributive Property over Addition* 👉 $a(b+c) = ab + ac$
2. *Distributive Property over Subtraction* 👉 $a(b-c) = ab - ac$

Formulas for Chapter: Equations and Inequalities

One-Step Equations

1. *Solving Addition Equations* 👉 $x + a = b \Rightarrow x = b - a$
2. *Solving Subtraction Equations* 👉 $x - a = b \Rightarrow x = b + a$
3. *Solving Multiplication Equations* 👉 $ax = b \Rightarrow x = \frac{b}{a}$
4. *Solving Division Equations* 👉 $\frac{x}{a} = b \Rightarrow x = ab$

Multi-Step Equations

1. *Combine Like Terms* 👉 $ax + bx = (a+b)x$
2. *Variables on One Side* 👉 $ax \pm c = bx \pm d \Rightarrow ax - bx = \pm(d-c)$
3. *Isolate Variable using Addition/Subtraction* 👉 $ax \pm b = c \Rightarrow ax = c \mp b$
4. *Isolate Variable using Multiplication/Division* 👉 $ax = b \Rightarrow x = \frac{b}{a}$

Multi-Step Inequalities

1. *Moving Variables to One Side* 👉 $ax \pm b \leq c \pm dx \Rightarrow (a \mp d)x \leq c \mp b$
2. *Isolate the Variable Using Multiplication/Division* 👉 $bx \leq d \Rightarrow x \leq \frac{d}{b}$, $b > 0$
3. *Reverse Inequality When Multiplying/Dividing by a Negative* 👉 $bx \leq d \Rightarrow x \geq \frac{d}{b}$, $b < 0$

📖 Formulas For Chapter: Lines and Slope

Finding Slope

1. *Slope from Two Points* 👉 Given two points $A(x_1, y_1)$ and $B(x_2, y_2)$, the slope of the line through A and B is: slope $= \frac{y_2 - y_1}{x_2 - x_1}$

Graphing Lines Using Slope-Intercept Form

1. *Slope-Intercept Form of a Line* 👉 $y = mx + b$ where m is the slope and b is the y-intercept.
2. *Finding the y-Intercept* 👉 Set $x = 0$ in the equation $y = mx + b$ to find the y-intercept $(0, b)$.

Finding Midpoint

☞ *Midpoint Formula* — $M = \left(\frac{x_1+x_2}{2}, \frac{y_1+y_2}{2}\right)$ for endpoints $A(x_1, y_1)$ and $B(x_2, y_2)$.

Finding Distance of Two Points

☞ *Distance Formula* — $d = \sqrt{(x_2-x_1)^2 + (y_2-y_1)^2}$

Formulas For Chapter Polynomials

Simplifying Polynomials

☞ *General Form of a Polynomial* — $P(x) = a_n x^n + a_{n-1} x^{n-1} + \ldots + a_1 x + a_0$

Adding and Subtracting Polynomials

☞ *Addition of Polynomials* — $(a_n x^n + a_{n-1} x^{n-1} + \cdots + a_1 x + a_0) + (b_n x^n + b_{n-1} x^{n-1} + \cdots + b_1 x + b_0) = (a_n + b_n)x^n + (a_{n-1} + b_{n-1})x^{n-1} + \cdots + (a_1 + b_1)x + (a_0 + b_0)$

☞ *Subtraction of Polynomials* — $(a_n x^n + a_{n-1} x^{n-1} + \cdots + a_1 x + a_0) - (b_n x^n + b_{n-1} x^{n-1} + \cdots + b_1 x + b_0) = (a_n - b_n)x^n + (a_{n-1} - b_{n-1})x^{n-1} + \cdots + (a_1 - b_1)x + (a_0 - b_0)$

Multiplying and Dividing Monomials

☞ *Multiplication of Monomials* — $c_1 x^a y^b \times c_2 x^m y^n = (c_1 \times c_2) x^{a+m} y^{b+n}$

☞ *Division of Monomials* — $\frac{c_1 x^a y^b}{c_2 x^m y^n} = \left(\frac{c_1}{c_2}\right) x^{a-m} y^{b-n}$ where $c_2 \neq 0$

Multiplying Binomials

☞ *The FOIL Method* — $(x+a)(x+b) = x^2 + bx + ax + ab = x^2 + (a+b)x + ab$.

Factoring Trinomials

1. *Reverse FOIL Method* 👉 $x^2 + (a+b)x + ab = (x+a)(x+b)$
2. *Difference of Squares* 👉 $a^2 - b^2 = (a+b)(a-b)$
3. *Perfect Square Trinomial (Sum)* 👉 $a^2 + 2ab + b^2 = (a+b)^2$
4. *Perfect Square Trinomial (Difference)* 👉 $a^2 - 2ab + b^2 = (a-b)^2$

📖 Formulas For Chapter: Geometry and Solid Figures

Complementary and Supplementary angles

1. *Complementary Angle Formula* 👉 If x and y are complementary, then $x + y = 90°$.
2. *Supplementary Angle Formula* 👉 If x and y are supplementary, then $x + y = 180°$.

Triangles

1. *Angle Sum Property* 👉 In any triangle, the sum of the interior angles is always 180°: $A + B + C = 180°$.
2. *Area of a Triangle* 👉 The area A of a triangle with base b and height h is given by: $A = \frac{1}{2} \times b \times h$.

The Pythagorean Theorem

1. *Pythagorean Theorem* 👉 In a right-angled triangle, the square of the length of the hypotenuse (c) is equal to the sum of the squares of the other two sides (a and b): $a^2 + b^2 = c^2$

Special Right Triangles

1. *45° − 45° − 90° Triangle Side Ratios* 👉 For 45° − 45° − 90° triangle: sides are $a, a, a\sqrt{2}$
2. *30° − 60° − 90° Triangle Side Ratios* 👉 For 30° − 60° − 90° triangle: sides are $a, a\sqrt{3}, 2a$

Polygons

1. Perimeter of a Square ☞ $P_{\text{square}} = 4s$
2. Area of a Square ☞ $A_{\text{square}} = s^2$
3. Perimeter of a Rectangle ☞ $P_{\text{rectangle}} = 2(\text{width} + \text{length})$
4. Area of a Rectangle ☞ $A_{\text{rectangle}} = \text{width} \times \text{length}$
5. Perimeter of a Trapezoid ☞ $P_{\text{trapezoid}} = \text{base}_1 + \text{base}_2 + \text{leg}_1 + \text{leg}_2$
6. Area of a Trapezoid ☞ $A_{\text{trapezoid}} = \frac{1}{2} \times (\text{base}_1 + \text{base}_2) \times \text{height}$
7. Perimeter of a Hexagon ☞ $P_{\text{hexagon}} = 6s$
8. Area of a Regular Hexagon ☞ $A_{\text{hexagon}} = \frac{3\sqrt{3}}{2} \times s^2$
9. Perimeter of a Parallelogram ☞ $P_{\text{parallelogram}} = 2(a+b)$
10. Area of a Parallelogram ☞ $A_{\text{parallelogram}} = \text{base} \times \text{height}$

Circles

1. Diameter and Radius ☞ $d = 2r$ where d is the diameter, and r is the radius of the circle.
2. Circumference of a Circle ☞ $C = 2\pi r$ where C is the circumference, r is the radius, and $\pi \approx 3.14$.
3. Area of a Circle ☞ $A = \pi r^2$ where A is the area, r is the radius, and $\pi \approx 3.14$.

Cubes

1. Volume of a Cube ☞ $V = a^3$ where a is the length of a side of the cube.
2. Surface Area of a Cube ☞ $A_s = 6a^2$ where a is the length of a side of the cube.

Rectangular Prisms

1. Volume of a Rectangular Prism ☞ $V = l \times w \times h$
2. Surface Area of a Rectangular Prism ☞ $A = 2 \times (wh + lw + lh)$

Cylinder

1. *Volume of a Cylinder* ☞ $V = \pi r^2 h$ where V is the volume, r is the radius, and h is the height of the cylinder.

2. *Surface Area of a Cylinder* ☞ $A = 2\pi r(r+h)$ where A is the surface area, r is the radius, and h is the height of the cylinder.

Formulas For Chapter: Statistics

Mean, Median, Mode, and Range of the Given Data

1. *Calculating Mean* ☞ Mean $= \frac{x_1 + x_2 + \cdots + x_n}{n}$
2. *Determining Mode* ☞ Mode is the value(s) that occurs most frequently in the data set.
3. *Determining Median (Odd count of numbers)* ☞ Median $= x_{\frac{n+1}{2}}$
4. *Determining Median (Even count of numbers)* ☞ Median $= \frac{x_{\frac{n}{2}} + x_{\frac{n}{2}+1}}{2}$
5. *Calculating Range* ☞ Range $=$ Largest data value $-$ Smallest data value

Pie Graph

1. *Calculate Percentage of Category* ☞ Percentage $= \left(\frac{\text{Number of instances in category}}{\text{Total number of instances}}\right) \times 100\%$
2. *Calculate Angle of Sector Using Degrees* ☞ Angle in degrees $= \left(\frac{\text{Number of instances in category}}{\text{Total number of instances}}\right) \times 360°$
3. *Calculate Quantity from Percentage* ☞ Quantity $=$ Percentage $\times$ Total quantity

Probability Problems

1. *Basic Probability Formula* ☞ $P(E) = \frac{\text{number of desired outcomes}}{\text{number of total outcomes}}$

Permutations and Combinations

1. *Factorial Definition* 👉 $n! = n \times (n-1) \times (n-2) \times \cdots \times 2 \times 1$ for $n \geq 1$, with $0! = 1$.
2. *Permutations Formula* 👉 $P(n,r) = \frac{n!}{(n-r)!}$ for selecting r different items out of n where order matters.
3. *Combinations Formula* 👉 $C(n,r) = \frac{n!}{r!(n-r)!}$ for selecting r different items out of n where order does not matter.
4. *Permutations with Repetition* 👉 n^r where n is the number of choices for each position and r is the total number of positions.

Formulas For Chapter: Functions Operations

Function Notation and Evaluation

1. *Function Notation* 👉 $f(x)$ represents the function value at x.
2. *Evaluating a Function* 👉 To evaluate $f(x)$ at $x = a$, replace x with a: $f(a) = f(x)|_{x=a}$.

Adding and Subtracting Functions

1. *Adding Functions* 👉 $(f+g)(x) = f(x) + g(x)$
2. *Subtracting Functions* 👉 $(f-g)(x) = f(x) - g(x)$

Multiplying and Dividing Functions

1. *Multiplication of Functions* 👉 $(f \cdot g)(x) = f(x) \cdot g(x)$
2. *Division of Functions* 👉 $\left(\frac{f}{g}\right)(x) = \frac{f(x)}{g(x)}$, provided $g(x) \neq 0$

Composition of Functions

1. *Composition of Functions* 👉 $(f \circ g)(x) = f(g(x))$

It is Time to Test Yourself

It's time to refine your skills with a practice examination designed to simulate the ASVAB Math Test. Engaging with the practice tests will help you to familiarize yourself with the test format and timing, allowing for a more effective test day experience. After completing a test, use the provided answer key to score your work and identify areas for improvement.

Before You Start

To make the most of your practice test experience, please ensure you have:
- A pencil for marking answers on the answer sheet.
- A timer to manage pacing, replicating potential time constraints in other testing scenarios.

Please note the following important points as you prepare to take your practice test:
- It's okay to guess! There is no penalty for incorrect answers, so make sure to answer every question.
- After completing the test, review the answer key to understand any mistakes. This review is crucial for your learning and preparation.
- An answer sheet is provided for you to record your answers. Make sure to use it.
- For each multiple-choice question, you will be presented with possible choices. Your task is to choose the best one.

Good Luck! Your preparation and practice are the keys to success.

16. Practice Test 1

ASVAB Math Practice Test

Section 1: Arithmetic Reasoning

Total time for this section: 36 Minutes

30 questions

You may NOT use a calculator on this section

16.1 Practices

1) Lila was contracted to conduct four identical workshops, each requiring 12 hours of work. She earns $30 per hour. What is Lila's total earnings for conducting all workshops?

☐ A. $3600

☐ B. $1440

☐ C. $4800

☐ D. $2430

2) Max is 7 years older than his sister Sophie, and Sophie is 5 years younger than their brother Ethan. If the

16.1 Practices

sum of their ages is 99, how old is Sophie?

- ☐ A. 31
- ☐ B. 28
- ☐ C. 29
- ☐ D. 25

3) Alex is planning a trip to a destination 750 miles away. If he travels at an average speed of 60 mph, how long will his round-trip journey take?

- ☐ A. 600 Minutes
- ☐ B. 750 Minutes
- ☐ C. 1500 Minutes
- ☐ D. 1800 Minutes

4) Mike distributes 10 comic books to each of his friends. If Mike gives away all of his comic books, which number could be the total number he distributed?

- ☐ A. 130
- ☐ B. 155
- ☐ C. 178
- ☐ D. 192

5) A jar contains only yellow and green marbles. The probability of selecting a yellow marble at random is one third. If there are 90 green marbles, how many marbles are there in total?

- ☐ A. 100
- ☐ B. 135
- ☐ C. 150
- ☐ D. 180

6) You are tasked with recording the hourly snowfall over a 6-hour period to calculate the average. The measurements are as follows: 6 am: 3 inches, 9 am: 25 inches, 10 am: 30 inches, 11 am: 35 inches, 12 pm: 38 inches, 1 pm: 40 inches. What is the average snowfall?

- ☐ A. 28.5 inches
- ☐ B. 28.8 inches

☐ C. 28.3 inches

☐ D. 29.1 inches

7) A local library charges an annual membership fee of $20, with an additional $0.25 per book for the first 50 books borrowed, and $0.15 for every book thereafter. What is the total cost for a member who borrows 70 books in a year?

☐ A. $30

☐ B. $33.50

☐ C. $35.50

☐ D. $42

8) A delivery vehicle travels 40 miles on Monday, 52 miles on Tuesday, and 55 miles on Thursday. What is the average distance traveled per day?

☐ A. 45 Miles

☐ B. 49 Miles

☐ C. 51 Miles

☐ D. 53 Miles

9) Four friends contribute $12.25, $14.75, $16.50, and $20.00 respectively for a joint birthday gift. What is the total amount they can spend on the gift?

☐ A. $58.50

☐ B. $62.00

☐ C. $63.50

☐ D. $65.75

10) A landscaper needs to transport soil weighing a total of 18,000 pounds using a van with a maximum capacity of 4,500 pounds per trip. How many trips will the landscaper need to transport all the soil?

☐ A. 3 Trips

☐ B. 4 Trips

☐ C. 5 Trips

☐ D. 6 Trips

16.1 Practices

11) A security guard checks the surveillance cameras every 75 minutes. In a 10-hour shift, how many times will the surveillance be checked?

☐ A. 6 Times

☐ B. 7 Times

☐ C. 8 Times

☐ D. 9 Times

12) A school has 20 dozens of books. After giving away 70 books to a charity, how many books does the school have left?

☐ A. 190

☐ B. 170

☐ C. 150

☐ D. 120

13) In a deck of cards, there are 5 spades, 4 hearts, 8 clubs, and 9 diamonds. What is the probability of randomly picking a club?

☐ A. $\frac{4}{13}$

☐ B. $\frac{8}{13}$

☐ C. $\frac{1}{8}$

☐ D. $\frac{1}{3}$

14) What is the prime factorization of 450?

☐ A. $2 \times 3 \times 3 \times 5 \times 5$

☐ B. $2 \times 5 \times 5 \times 9$

☐ C. $3 \times 3 \times 5 \times 10$

☐ D. $2 \times 2 \times 3 \times 3 \times 5$

15) A gambler starts with $200 at a casino. She loses $60 on a slot machine and then $70 on a card game. How much money does she have left?

☐ A. $60

☐ B. $70

☐ C. $80

☐ D. $90

16) A pet care service finds that 4 workers can walk 12 dogs. How many dogs can 6 workers walk at the same ratio?

☐ A. 16

☐ B. 18

☐ C. 20

☐ D. 24

17) If $\frac{3y}{2x} - \frac{y}{4x} = \frac{(...)}{4x}$ and $x \neq 0$, what expression is represented by $(...)$?

☐ A. $2y$

☐ B. $6y$

☐ C. $5y$

☐ D. $8y$

18) A restaurant offers 4 choices of appetizers, 5 choices of main courses, and 3 choices of desserts. How many different three-course meals are possible?

☐ A. 15

☐ B. 30

☐ C. 60

☐ D. 120

19) In a sequence, each number is 3 less than triple the number that comes before it. If 54 is a number in the sequence, what number comes just before it?

☐ A. 17

☐ B. 19

☐ C. 21

☐ D. 23

20) The average of four numbers is 22. If a fifth number, 36, is added, what is the new average?

☐ A. 23.2

☐ B. 22.5

☐ C. 25.2

16.1 Practices

☐ D. 24.8

21) A laboratory sample has a mass of 125 milligrams. What is the sample's mass in grams?

☐ A. 0.125

☐ B. 0.0125

☐ C. 1.25

☐ D. 12.5

22) How many hours are there in 2,160 minutes?

☐ A. 24 hours

☐ B. 30 hours

☐ C. 36 hours

☐ D. 40 hours

23) Consider the set of numbers $\{3, 6, 7, 9, 14, 15\}$. Removing which number will change the average to 8?

☐ A. 3

☐ B. 6

☐ C. 10

☐ D. 14

24) A backpack originally priced at $60.00 was on sale for 20% off. Maria received an additional 10% student discount on the sale price. How much did Maria pay for the backpack?

☐ A. $43.20

☐ B. $46.80

☐ C. $48.00

☐ D. $52.80

25) In a music class, there are 15 boys and 10 girls. What is the ratio of the number of boys to the number of girls?

☐ A. 1:2

☐ B. 1:1.5

☐ C. 3:2

☐ D. 2:1

26) Which of the following expressions is not equal to 8?

- A. $32 \times \frac{1}{4}$
- B. $4 \times \frac{8}{4}$
- C. $7 \times \frac{8}{7}$
- D. $8 \times \frac{1}{8}$

27) Which of the following is greater than $\frac{15}{10}$?

- A. $\frac{1}{3}$
- B. $\frac{7}{3}$
- C. $\frac{2}{5}$
- D. 1

28) Which of the following is the same as 0.0000000000000052143?

- A. 5.2143×10^{14}
- B. 5.2143×10^{-14}
- C. $52,143 \times 10^{-10}$
- D. 52.143×10^{-13}

29) Ron spent $45 for a jacket. This was $15 more than double what he spent for a hat. How much was the hat?

- A. $12
- B. $15
- C. $18
- D. $20

30) What is the value of the following expression? $|-7| + 10 \times 2\frac{1}{2} + (-4)^2$?

- A. 33
- B. 39
- C. 48
- D. 52

ASVAB Math Practice Test

Section 2: Mathematics Knowledge

Total time for this section: 24 Minutes

25 questions

You may NOT use a calculator on this section

1) If $x = 5$, what is the value of y in this equation? $y = \frac{x^3}{5} - 4$

   ☐ A. 22

   ☐ B. 21

   ☐ C. 18

   ☐ D. 16

2) The tenth root of 1024 is:

   ☐ A. 6

   ☐ B. 5

   ☐ C. 4

   ☐ D. 2

3) A circle has a diameter of 10 inches. What is its approximate area? ($\pi = 3.14$)

   ☐ A. 157 square inches

   ☐ B. 78.5 square inches

   ☐ C. 31.4 square inches

   ☐ D. 50 square inches

4) If $-16b = 128$, then $b =$ ____

   ☐ A. −8

- B. −4
- C. 8
- D. 4

5) In the diagram below, circle *P* represents the set of all prime numbers, circle *Q* represents the set of all positive numbers, and circle *R* represents the set of all multiples of 5. Which number could be replaced with *z*?

- A. 15
- B. 5
- C. −5
- D. −10

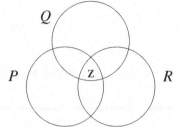

6) In the trapezoid shown below, what is the value of $p+q$?

- A. 180
- B. 230
- C. 250
- D. 360

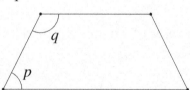

7) The shaded sector of a circle shown below has an area of 6π square feet. What is the circumference of the circle?

- A. 18π feet
- B. 16π feet
- C. 14π feet
- D. 12π feet

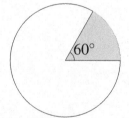

8) If $2^{30} = 2^{10} \times 2^y$, what is the value of *y*?

- A. 10
- B. 15
- C. 20
- D. 25

9) Which of the following is a right angle?

- A. 90°
- B. 45°
- C. 30°

16.1 Practices

☐ D. 60°

10) Factor this expression: $x^2 - 5x + 6$

☐ A. $(x-2)(x-3)$

☐ B. $(x+2)(x+3)$

☐ C. $(x-6)(x+1)$

☐ D. $(x+6)(x-6)$

11) Calculate the slope of the line that passes through the points $(4,-1)$ and $(-2,5)$.

☐ A. $\frac{2}{3}$

☐ B. -1

☐ C. 1

☐ D. -2

12) Compute the product of $\sqrt{49}$ and $\sqrt{81}$.

☐ A. 63

☐ B. 343

☐ C. 7

☐ D. 9

13) Anna drank 3 glasses of juice. The first glass contained $\frac{1}{6}$ liter, the second $\frac{2}{5}$ liter, and the third $\frac{1}{4}$ liter. What is the total volume of juice she drank?

☐ A. $\frac{17}{30}$ liter

☐ B. $\frac{49}{60}$ liter

☐ C. $\frac{11}{20}$ liter

☐ D. $\frac{3}{4}$ liter

14) Determine a factor of 63 other than 1 and 63.

☐ A. 3

☐ B. 8

☐ C. 10

☐ D. 15

15) The Greatest Common Factor (GCF) of 48 and a number y is 8, where y is greater than 10 but less than 50. How many possible values are there for y?

- ☐ A. 2
- ☐ B. 3
- ☐ C. 4
- ☐ D. 1

16) Evaluate: $50(2+0.2)^2 - 50$.

- ☐ A. 212
- ☐ B. 180
- ☐ C. 192
- ☐ D. 198

17) Find the number of integers between $\frac{9}{3}$ and $\frac{33}{4}$.

- ☐ A. 5
- ☐ B. 6
- ☐ C. 7
- ☐ D. 8

18) A rectangular pool holds 2,400 cubic feet of water. The pool is 20 feet long and 12 feet wide. What is the depth of the pool?

- ☐ A. 5 feet
- ☐ B. 8 feet
- ☐ C. 10 feet
- ☐ D. 12 feet

19) A student made a list of all possible products of 2 different numbers in the set $\{2, 3, 5, 7, 9\}$. What fraction of the products are even?

- ☐ A. $\frac{2}{5}$
- ☐ B. $\frac{1}{5}$
- ☐ C. $\frac{3}{5}$
- ☐ D. $\frac{4}{5}$

16.1 Practices

20) Solve the inequality:

$$|x+3| \geq 5$$

- ☐ A. $x \geq 2$ or $x \leq -8$
- ☐ B. $x \geq -2$ or $x \leq 8$
- ☐ C. $x \geq -8$ or $x \leq 2$
- ☐ D. $x \geq 8$ or $x \leq -2$

21) If $3m$ is a positive odd number, how many even numbers are in the range from $3m$ up to and including $3m+5$?

- ☐ A. 2
- ☐ B. 3
- ☐ C. 4
- ☐ D. 5

22) If $3x - 4y = 15$, what is x in terms of y?

- ☐ A. $x = \frac{4}{3}y + 5$
- ☐ B. $x = \frac{3}{4}y + 15$
- ☐ C. $x = -\frac{4}{3}y - 5$
- ☐ D. $x = -\frac{4}{3}y + 5$

23) In the triangle below, if the measure of angle A is 45 degrees, what is the value of z? (figure is NOT drawn to scale)

- ☐ A. 120
- ☐ B. 127
- ☐ C. 136
- ☐ D. 140

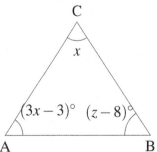

24) What is the perimeter of a circle that has an area of 64π square inches?

- ☐ A. 64π inches
- ☐ B. 32π inches
- ☐ C. 16π inches

☐ D. 8π inches

25) What are the zeros of the function: $f(x) = x^2 - 9x + 18$?

☐ A. 0

☐ B. 3, 6

☐ C. −3, −6

☐ D. 0, 3, 6

16.2 Answer Keys

 Section 1: Arithmetic Reasoning

1) B. $1440
2) C. 29
3) C. 1500 Minutes
4) A. 130
5) B. 135
6) A. 28.5 inches
7) C. $35.50
8) B. 49 Miles
9) C. $63.50
10) B. 4 Trips
11) C. 8 Times
12) B. 170
13) A. $\frac{4}{13}$
14) A. $2 \times 3 \times 3 \times 5 \times 5$
15) B. $70
16) B. 18
17) C. $5y$
18) C. 60
19) B. 19
20) D. 24.8
21) A. 0.125
22) C. 36 hours
23) D. 14
24) A. $43.20
25) C. 3:2
26) D. $8 \times \frac{1}{8}$
27) B. $\frac{7}{3}$
28) B. 5.2143×10^{-14}
29) B. $15
30) C. 48

 **Section 2: Mathematics Knowledge**

1) B. 21
2) D. 2
3) B. 78.5 square inches
4) A. −8
5) B. 5
6) A. 180
7) D. 12π feet
8) C. 20
9) A. 90°
10) A. $(x-2)(x-3)$
11) B. −1
12) A. 63
13) B. $\frac{49}{60}$ liter

14) A. 3
15) D. 1
16) C. 192
17) A. 5
18) C. 10 feet
19) A. $\frac{2}{5}$
20) A. $x \geq 2$ or $x \leq -8$
21) B. 3
22) A. $x = \frac{4}{3}y + 5$
23) B. 127
24) C. 16π inches
25) B. 3, 6

16.3 Answers with Explanation

 **Section 1: Arithmetic Reasoning**

1) Earnings for all workshops are calculated by multiplying the hourly rate by the number of hours:

$$\text{Earnings} = 12 \text{ hours} \times 4 \times \$30/\text{hour} = \$1440$$

So, Lila earns $1440 for all workshops, making option B correct.

2) Let Sophie's age be s. Max's age is $s+7$ and Ethan's age is $s+5$. The sum of their ages is:

$$s + (s+7) + (s+5) = 99 \rightarrow 3s + 12 = 99 \rightarrow 3s = 87 \rightarrow s = 29$$

Therefore, Sophie is 29 years old, making option C correct.

3) The round-trip distance is $2 \times 750 = 1500$ miles. The time taken is:

$$\text{Time} = \frac{1500 \text{ miles}}{60 \text{ mph}} = 25 \text{ hours} = 25 \times 60 \text{ minutes} = 1500 \text{ minutes}$$

Therefore, the journey takes 1500 minutes, making option C correct.

4) The total number of comic books distributed must be a multiple of 10. Among the options, only 130 is a multiple of 10. Therefore, option A is correct.

5) Let y be the number of yellow marbles. The total number of marbles is $y+90$. The probability of selecting a yellow marble is $\frac{y}{y+90} = \frac{1}{3}$:

$$\frac{y}{y+90} = \frac{1}{3} \rightarrow 3y = y + 90 \rightarrow 2y = 90 \rightarrow y = 45$$

The total number of marbles is $45 + 90 = 135$, making option B correct.

6) To find the average, sum up the snowfall and divide by the number of hours:

$$\text{Total Snowfall} = 3 + 25 + 30 + 35 + 38 + 40 = 171 \text{ inches}$$

$$\text{Average Snowfall} = \frac{171}{6} = 28.5 \text{ inches}$$

Therefore, the average snowfall is 28.5 inches, making option A correct.

7) The cost for the first 50 books is calculated as:

$$\text{Cost for 50 Books} = 0.25 \times 50 = \$12.50$$

The cost for the remaining 20 books is:

$$\text{Cost for 20 Books} = 0.15 \times 20 = \$3$$

The total cost is $20 + $12.50 + $3 = $35.50, making option C correct.

8) The total distance traveled is calculated as:

$$\text{Total Distance} = 40 + 52 + 55 = 147 \text{ miles}$$

$$\text{Average Distance} = \frac{147}{3} = 49 \text{ miles per day}$$

Therefore, the average distance is 49 miles per day, making option B correct.

9) Sum up the contributions:

$$\text{Total Contribution} = 12.25 + 14.75 + 16.50 + 20.00 = \$63.50$$

The maximum amount they can spend on the gift is $63.50, making option C correct.

10) The total weight of the soil is 18,000 pounds, and the van can carry 4,500 pounds per trip. To find out how

16.3 Answers with Explanation

many trips are required, divide the total weight by the capacity per trip:

$$\text{Number of trips} = \lceil \frac{18,000}{4,500} \rceil = \lceil 4 \rceil = 4.$$

The landscaper will need to make 4 trips to transport all the soil, making option B correct.

11) In a 10-hour shift (600 minutes), the number of checks is:

$$\text{Number of Checks} = \frac{600}{75} = 8$$

Therefore, the surveillance will be checked 8 times, making option C correct.

12) Initially, the school has $20 \times 12 = 240$ books. After donating 70 books:

$$\text{Books Left} = 240 - 70 = 170$$

Therefore, 170 books are left, making option B correct.

13) The total number of cards is $5 + 4 + 8 + 9 = 26$. The probability of picking a club is:

$$\text{Probability} = \frac{\text{Number of Clubs}}{\text{Total Number of Cards}} = \frac{8}{26} = \frac{4}{13}$$

Thus, the probability is $\frac{4}{13}$, making option A correct.

14) The prime factorization of 450 is:

$$450 = 2 \times 3 \times 3 \times 5 \times 5$$

Therefore, the correct answer is option A.

15) The gambler's remaining money is calculated by subtracting the losses from the starting amount:

$$200 - 60 - 70 = 70$$

She has $70 left, making option B correct.

16) Using the ratio of workers to dogs, calculate the number of dogs 6 workers can walk:

$$\frac{4 \text{ workers}}{12 \text{ dogs}} = \frac{6 \text{ workers}}{x \text{ dogs}} \rightarrow x = \frac{12 \times 6}{4} = 18$$

Thus, 6 workers can walk 18 dogs, making option B correct.

17) Simplify the equation to find the expression:

$$\frac{3y}{2x} - \frac{y}{4x} = \frac{5y}{4x}$$

The missing expression is $5y$, making option C correct.

18) Calculate the total combinations by multiplying the number of choices for each course:

$$4 \times 5 \times 3 = 60$$

There are 60 different meal combinations, making option C correct.

19) Solve for the previous number in the sequence:

$$3 \times \text{previous number} - 3 = 54 \rightarrow 3 \times \text{previous number} = 57 \rightarrow \text{previous number} = \frac{57}{3} = 19$$

The previous number is 19, making option B correct.

20) Calculate the new average by finding the total sum and dividing by the number of numbers:

$$\frac{4 \times 22 + 36}{5} = \frac{88 + 36}{5} = \frac{124}{5} = 24.8$$

The new average is 24.8, making option D correct.

21) Convert the mass from milligrams to grams by dividing by 1000:

$$\frac{125 \text{ mg}}{1000} = 0.125 \text{ g}$$

The mass of the sample in grams is 0.125, making option A correct.

16.3 Answers with Explanation

22) Convert the time from minutes to hours:

$$\frac{2160 \text{ minutes}}{60} = 36 \text{ hours}$$

There are 36 hours in 2160 minutes, making option C correct.

23) Determine which number, when removed, changes the average to 8:

$$\text{New total sum} = 8 \times 5 = 40 \rightarrow \text{Original sum} - \text{Removed number} = 40$$

Find the removed number:

$$3 + 6 + 7 + 9 + 14 + 15 - \text{Removed number} = 40$$

The removed number is 14, making option D correct.

24) Calculate the sale price after the first discount, then apply the student discount:

$$\text{Sale price} = 60.00 \times (1 - 0.20) = 48.00$$

$$\text{Final price} = 48.00 \times (1 - 0.10) = 43.20$$

Maria paid $43.20 for the backpack, making option A correct.

25) Determine the ratio of boys to girls:
$$\text{Ratio} = \frac{15}{10} = \frac{3}{2}$$

The ratio of boys to girls is 3:2, making option C correct.

26) Evaluate each expression to find which is not equal to 8:

$$A: 32 \times \frac{1}{4} = 8, \quad B: 4 \times \frac{8}{4} = 8, \quad C: 7 \times \frac{8}{7} = 8, \quad D: 8 \times \frac{1}{8} = 1$$

Expression D is not equal to 8, making option D correct.

27) Compare each fraction to $\frac{15}{10}$:

$$A: \frac{1}{3} < \frac{15}{10}, \quad B: \frac{7}{3} > \frac{15}{10}, \quad C: \frac{2}{5} < \frac{15}{10}, \quad D: 1 < \frac{15}{10}$$

The fraction greater than $\frac{15}{10}$ is $\frac{7}{3}$, making option B correct.

28) Express the number in scientific notation:

$$0.000000000000052143 = 5.2143 \times 10^{-14}$$

The correct representation in scientific notation is option B.

29) Let the price of the hat be x. The equation is:

$$45 = 2x + 15 \;\rightarrow\; 30 = 2x \;\rightarrow\; x = 15$$

The price of the hat was $15, making option B correct.

30) Calculate the expression:

$$|-7| + 10 \times 2\frac{1}{2} + (-4)^2 = 7 + 25 + 16 = 48$$

The value of the expression is 48, making option C correct.

16.3 Answers with Explanation

 ## Section 2: Mathematics Knowledge

1) Substitute $x = 5$ into the equation:

$$y = \frac{5^3}{5} - 4 = \frac{125}{5} - 4 = 25 - 4 = 21$$

The value of y is 21, making option B correct.

2) Calculate the tenth root of 1024:

$$\sqrt[10]{1024} = \sqrt[10]{2^{10}} = 2$$

The tenth root of 1024 is 2, making option D correct.

3) The radius is half of the diameter, so $r = 5$ inches. Calculate the area:

$$A = \pi r^2 = 3.14 \times 5^2 = 3.14 \times 25 = 78.5$$

The area of the circle is 78.5 square inches, making option B correct.

4) Solve for b:

$$-16b = 128 \rightarrow b = \frac{128}{-16} = -8$$

The value of b is -8, making option A correct.

5) 5 is a prime number, positive, and a multiple of 5, fitting all three sets. The number that can replace z is 5, making option B correct.

6) In any trapezoid, the sum of the interior angles adjacent to any one side is 180 degrees:

$$p + q = 180°$$

Therefore, the sum of $p + q$ is 180°, making option A correct.

7) To find the circumference, we first find the radius using the area of the sector. For a full circle, the area is

πr^2, so for our sector:

$$6\pi = \frac{1}{6}\pi r^2 \rightarrow r^2 = 36 \rightarrow r = \sqrt{36} = 6 \text{ feet}$$

The circumference is $2\pi r$, $2\pi \times 6 = 12\pi$, option D.

8) Apply the law of exponents:

$$2^{30} = 2^{10} \times 2^y = 2^{10+y} \rightarrow 30 = 10 + y \rightarrow y = 20$$

The value of y is 20, making option C correct.

9) A right angle is exactly $90°$. Therefore, option A is correct.

10) Factor by looking for two numbers that multiply to 6 and add up to -5:

$$x^2 - 5x + 6 = (x-2)(x-3)$$

Therefore, option A is the correct factored form.

11) The slope formula is $m = \frac{y_2 - y_1}{x_2 - x_1}$. Plugging in the points $(4, -1)$ and $(-2, 5)$:

$$m = \frac{5 - (-1)}{-2 - 4} = \frac{6}{-6} = -1$$

The slope of the line is -1, making option B correct.

12) The square roots of 49 and 81 are 7 and 9 respectively. Multiplying these values:

$$\sqrt{49} \times \sqrt{81} = 7 \times 9 = 63$$

The product is 63, making option A correct.

13) Sum the fractions:

$$\frac{1}{6} + \frac{2}{5} + \frac{1}{4} = \frac{10}{60} + \frac{24}{60} + \frac{15}{60} = \frac{49}{60}$$

Anna drank a total of $\frac{49}{60}$ liter, making option B correct.

16.3 Answers with Explanation

14) Among the options, 3 is a factor of 63:

$$63 \div 3 = 21$$

Since 63 can be divided evenly by 3, 3 is a factor of 63, making option A correct.

15) Considering y must be greater than 10 and less than 50, the possible values for y (with GCF 8) are:

$$16, 24, 32, 40, 48.$$

By checking these numbers, we find that only the number 40 has the required condition. Therefore, 1 possible value, making option D correct.

16) First, calculate $(2+0.2)^2$:

$$(2.2)^2 = 4.84$$

Then, substitute into the expression:

$$50(4.84) - 50 = 242 - 50 = 192$$

The value of the expression is 192, making option C correct.

17) The integers between $\frac{9}{3} = 3$ and $\frac{33}{4} = 8.25$ are $4, 5, 6, 7, 8$. There are 5 integers, making option A correct.

18) The volume of the pool is given by length $\times$ width $\times$ depth. Rearranging to find depth:

$$\text{Depth} = \frac{\text{Volume}}{\text{Length} \times \text{Width}} = \frac{2,400}{20 \times 12} = \frac{2,400}{240} = 10$$

The depth of the pool is 10 feet, making option C correct.

19) First, let's calculate the total number of distinct products. Since there are 5 numbers in the set and we need to choose 2 different numbers, the total number of combinations is:

$$C(5,2) = \frac{5!}{2! \times (5-2)!} = 10$$

For a product to be even, one of the factors must be even. In this set, only the number 2 is even. The combinations that include 2 are:

$$2 \times 3, 2 \times 5, 2 \times 7, 2 \times 9.$$

There are 4 such combinations. Therefore, the fraction of the products that are even is: $\frac{4}{10} = \frac{2}{5}$. Therefore the correct answer is A.

20) Break the absolute value inequality into two cases:

$$x+3 \geq 5 \rightarrow x \geq 2$$

and

$$x+3 \leq -5 \rightarrow x \leq -8$$

Therefore, $x \geq 2$ or $x \leq -8$, making option A correct.

21) The range from $3m$ to $3m+5$ (including $3m+5$) includes five numbers. Since $3m$ is odd, the even numbers in this range will be $3m+1$, $3m+3$, and $3m+5$. Therefore, there are 3 even numbers, making option B correct.

22) To solve for x, rearrange the equation:

$$3x = 4y+15 \rightarrow x = \frac{4y+15}{3} = \frac{4}{3}y+5$$

Thus, $x = \frac{4}{3}y+5$, making option A correct.

23) The angle A is $45°$:

$$3x-3 = 45 \rightarrow 3x = 48 \rightarrow x = 16.$$

Since the sum of angles in a triangle is 180 degrees:

$$x+3x-3+z-8 = 180 \rightarrow 16+3(16)-3+z-8 = 180 \rightarrow z = 127°.$$

Therefore, the value of z is $127°$, making option B correct.

16.3 Answers with Explanation

24) First, find the radius of the circle:

$$\text{Area} = \pi \times \text{radius}^2 \rightarrow 64\pi = \pi \times \text{radius}^2 \rightarrow \text{radius} = 8 \text{ inches}$$

The perimeter of a circle is its diameter times the π number:

$$\text{Perimeter} = 2(\text{radius}) \times \pi = 2 \times 8 \times \pi = 16\pi \text{ inches}$$

The perimeter is 16π inches, making option C correct.

25) To find the zeros, set the function equal to 0:

$$x^2 - 9x + 18 = 0$$

Factoring, we get:

$$(x-3)(x-6) = 0$$

Therefore, the zeros are $x = 3$ and $x = 6$, making option B correct.

17. Practice Test 2

ASVAB Math Practice Test

Section 1: Arithmetic Reasoning

Total time for this section: 36 Minutes

30 questions

You may NOT use a calculator on this section

17.1 Practices

1) Susan has been dedicating 8 hours a day to a coding project, 5 days a week. If this schedule is maintained for 3 weeks, how many hours in total has Susan worked on the project?

☐ A. 120

☐ B. 150

☐ C. 240

☐ D. 360

2) Set X consists of all integers from 20 to 200, inclusive, while set Y contains all integers from 100 to 220,

17.1 Practices

inclusive. Determine the number of integers that are in set X but not in set Y.

☐ A. 70

☐ B. 79

☐ C. 80

☐ D. 89

3) In a class of 80 students, 48 are male. Calculate the percentage of the class that is female.

☐ A. 35%

☐ B. 40%

☐ C. 50%

☐ D. 60%

4) Thomas spends $100 at a restaurant. If the tax rate is 10% of the meal price, and he leaves a 20% tip on the total bill including tax, what is his total expenditure?

☐ A. $121

☐ B. $132

☐ C. $144

☐ D. $154

5) During a week of training, Lily records her running times as follows: 70, 65, 55, 60, 65, and 63 seconds. What is the approximate average of her three fastest times?

☐ A. 56 seconds

☐ B. 59 seconds

☐ C. 61 seconds

☐ D. 64 seconds

6) To carpet a rectangular hall that measures 18 feet by 20 feet, how many square feet of carpet is required?

☐ A. 360 square feet

☐ B. 380 square feet

☐ C. 400 square feet

☐ D. 440 square feet

7) What number should 2.54821 be multiplied by to get 2548.21?

- [ ] A. 1,000
- [ ] B. 10,000
- [ ] C. 100,000
- [ ] D. 1,000,000

8) Identify which of the following is NOT a factor of 100:

- [ ] A. 4
- [ ] B. 5
- [ ] C. 25
- [ ] D. 14

9) John earns $28.00 an hour at a clinic. After a 5% salary increase, what will be his new hourly rate?

- [ ] A. $29.40 an hour
- [ ] B. $30.00 an hour
- [ ] C. $31.20 an hour
- [ ] D. $33.60 an hour

10) In a photography contest, David and Sarah took the same number of photos. David took three times as many photos as Alice. If Sarah took 18 more photos than Alice, how many photos did Alice take?

- [ ] A. 6
- [ ] B. 9
- [ ] C. 12
- [ ] D. 15

11) Julia is 5 feet 9 inches tall, and Eva is 4 feet 10 inches tall. What is the difference in height, in inches, between Julia and Eva?

- [ ] A. 8 inches
- [ ] B. 11 inches
- [ ] C. 13 inches
- [ ] D. 15 inches

12) Calculate the average of the numbers: 18, 26, 15, and 21.

- [ ] A. 18.5

17.1 Practices

☐ B. 20

☐ C. 21.5

☐ D. 22.3

13) A rectangular garden measures 120 feet in length and 60 feet in width. How far, in yards, does one walk if they circle the garden once?

☐ A. 150 yards

☐ B. 170 yards

☐ C. 190 yards

☐ D. 120 yards

14) If a car travels for 2.5 hours at an average speed of 20 miles per hour, what is the total distance covered?

☐ A. 50 miles

☐ B. 45 miles

☐ C. 40 miles

☐ D. 35 miles

15) The sum of 5 numbers is between 75 and 100. What could be the average (arithmetic mean) of these numbers?

☐ A. 14

☐ B. 13

☐ C. 19

☐ D. 21

16) A chef prepares an average of 12 dishes per hour. At this rate, how many hours will it take to prepare 1,200 dishes?

☐ A. 90 hours

☐ B. 100 hours

☐ C. 110 hours

☐ D. 120 hours

17) There are 100 books that need to be organized and 10 volunteers available. If 10 books remain unorganized by the end of the day, what is the average number of books that each volunteer has organized?

☐ A. 8
☐ B. 9
☐ C. 10
☐ D. 11

18) A gardener was making $8.00 per hour and received a raise to $8.40 per hour. What percentage increase was the gardener's raise?

☐ A. 3%
☐ B. 4%
☐ C. 5%
☐ D. 7%

19) A map uses a scale of 1 inch to represent 3 miles. What is the actual distance represented by 5.5 inches?

☐ A. 15 miles
☐ B. 16.5 miles
☐ C. 17 miles
☐ D. 18 miles

20) A vending machine accepts only dimes and quarters. A snack costs 50, a drink costs 75, and a pack of gum costs 25 dimes. How many dimes are needed to buy one snack, two drinks, and two pack of gum?

☐ A. 16 dimes
☐ B. 18 dimes
☐ C. 20 dimes
☐ D. 25 dimes

21) A clock's second hand rotates 360 degrees every minute. How many complete rotations does the second hand make in 6 hours?

☐ A. 180
☐ B. 360
☐ C. 720
☐ D. 1,080

22) What is the product of the square root of 64 and the square root of 49?

17.1 Practices

- ☐ A. 28
- ☐ B. 32
- ☐ C. 56
- ☐ D. 64

23) The least of 5 consecutive even integers is p, and the greatest is q. What is the value of $\frac{p+q}{2}$ in terms of p?

- ☐ A. $p+4$
- ☐ B. $p+8$
- ☐ C. $p+6$
- ☐ D. $2p+4$

24) A cake recipe calls for $3\frac{1}{4}$ cups of sugar. If you only have $2\frac{1}{2}$ cups of sugar, how much more sugar is needed?

- ☐ A. $\frac{3}{4}$
- ☐ B. $\frac{5}{8}$
- ☐ C. $\frac{7}{8}$
- ☐ D. $\frac{1}{2}$

25) Convert 0.087 to a percent.

- ☐ A. 0.87%
- ☐ B. 8.7%
- ☐ C. 87%
- ☐ D. 870%

26) $-24 + 8 \times (-3) - [6 + 18 \times (-3)] \div 3 =$

- ☐ A. -12
- ☐ B. -36
- ☐ C. -60
- ☐ D. -32

27) A basketball team has $25,000 to spend on equipment. They spent $18,000 on uniforms. The cost of each basketball is $150. What inequality represents the number of basketballs the team can purchase?

- ☐ A. $150x + 18,000 \leq 25,000$

☐ B. $150x + 18,000 \geq 25,000$

☐ C. $18,000x + 150 \leq 25,000$

☐ D. $18,000x + 150 \geq 25,000$

28) What is the value of x in the following equation? $\frac{x+10}{x} = \frac{8}{4}$

☐ A. $\frac{1}{10}$

☐ B. 10

☐ C. 40

☐ D. 20

29) If $m = 8$ and $n = -4$, what is the value of $\frac{7-9(4+n)}{4m-7(3-n)} =$?

☐ A. $-\frac{3}{11}$

☐ B. $-\frac{4}{9}$

☐ C. $\frac{5}{14}$

☐ D. $-\frac{7}{17}$

30) Leo scored an average of 75 per test in his first 3 tests. In his 4th test, he scored 85. What was Leo's mean score for the 4 tests?

☐ A. 70.5

☐ B. 72

☐ C. 77.5

☐ D. 80.2

ASVAB Math Practice Test
Section 2: Mathematics Knowledge

Total time for this section: 24 Minutes

25 questions

You may NOT use a calculator on this section

1) $(2x+8)(3x-6) = ?$

   ☐ A. $6x^2 + 12x - 48$

   ☐ B. $6x^2 + 8x - 12$

   ☐ C. $6x^2 + 14x + 24$

   ☐ D. $6x^2 - 14x - 48$

2) In the infinitely repeating decimal of $\frac{1}{13} = 0.\overline{076923}$, what is the 73rd digit?

   ☐ A. 6

   ☐ B. 9

   ☐ C. 7

   ☐ D. 0

3) What is the perimeter of a triangle with sides measuring 5, 7, and 8 units?

   ☐ A. 15 units

   ☐ B. 20 units

   ☐ C. 21 units

   ☐ D. 25 units

4) If x is a positive integer divisible by 7, and $x < 100$, what is the greatest possible value of x?

   ☐ A. 91

☐ B. 93
☐ C. 98
☐ D. 99

5) There are two juicers in a juice bar. Juicer 1 produces three times as much juice as juicer 2. If the juice bar produced a total of 20 liters of juice on Monday, how many liters did juicer 2 produce?

☐ A. 5
☐ B. 10
☐ C. 15
☐ D. 20

6) Yesterday Kylie completed 10% of her essay and today she finished a 20% of the reminder. What percentage of the essay is still unfinished?

☐ A. 70%
☐ B. 60%
☐ C. 72%
☐ D. 82%

7) If $10 + 3x$ is 19 more than 24, what is the value of $5x$?

☐ A. 32
☐ B. 55
☐ C. 50
☐ D. 60

8) In a box of red and green marbles, the ratio of red marbles to green marbles is 3 : 4. If the box contains 12 green marbles, how many red marbles are there?

☐ A. 6
☐ B. 15
☐ C. 12
☐ D. 9

9) In the xy−plane, the point $(5, 4)$ and $(2, 1)$ are on line l. Which of the following equations of lines is parallel to line l?

17.1 Practices

☐ A. $y = 2x - 1$

☐ B. $y = x + 1$

☐ C. $y = -x + 3$

☐ D. $y = -2x - 4$

10) How many positive odd factors of 90 are greater than 5 and less than 30?

☐ A. 0

☐ B. 3

☐ C. 2

☐ D. 4

11) If $y = (2x^4)^2$, which of the following expressions is equal to y?

☐ A. $4x^8$

☐ B. $8x^8$

☐ C. $4x^{16}$

☐ D. $16x^6$

12) The equation of a line is given as: $y = 4x + 5$. Which of the following points does not lie on the line?

☐ A. (1, 9)

☐ B. (0, 5)

☐ C. (−1, 2)

☐ D. (2, 13)

13) When $M + N = 15$ and $4S + N = 20$, what is the value of S?

☐ A. 3

☐ B. 2.5

☐ C. 2

☐ D. It cannot be determined from the information given.

14) What is the distance between the points (4, 6) and (1, 2)?

☐ A. 4

☐ B. 5

☐ C. 6

☐ D. 7

15) $x^2 - 49 = 0$, x could be:

☐ A. 7

☐ B. 8

☐ C. 9

☐ D. 10

16) A garden is shaped as a rectangle and measures 140 feet by 180 feet. What is its total area?

☐ A. 25,200 square feet

☐ B. 5,040 square feet

☐ C. 2,520 square feet

☐ D. 3,180 square feet

17) In the figure below, line C is parallel to line D. If angle y is 65 degrees, what is the value of angle x?

☐ A. 25 degrees

☐ B. 65 degrees

☐ C. 115 degrees

☐ D. 155 degrees

18) In a parallelogram, the ratio of two adjacent sides is 3 : 4. If the perimeter is 56 cm, find the lengths of its sides.

☐ A. 12 cm, 16 cm

☐ B. 9 cm, 12 cm

☐ C. 15 cm, 21 cm

☐ D. 18 cm, 24 cm

19) Max bought a tree that is initially 10 inches tall. The tree grows 5 inches every year. If Max's tree's height is expressed as a function of time, what does the y-intercept represent?

☐ A. The y-intercept represents the rate of growth of the tree which is 5 inches.

☐ B. The y-intercept represents the starting height of 10 inches.

☐ C. The y-intercept represents the starting height of 5 inches.

17.1 Practices

☐ D. The y-intercept represents the rate of growth of tree which is 4 inches per year.

20) One fourth of the cube of 6 is:

☐ A. 54

☐ B. 81

☐ C. 108

☐ D. 216

21) Calculate the sum of the prime numbers in this set:

$$18, 15, 17, 21, 23, 26, 31, 42, 37$$

☐ A. 58

☐ B. 71

☐ C. 108

☐ D. 125

22) Consider P is the midpoint of a line segment QR. If $QR = 4y - 2x$ cm, what is the length of segment PR?

☐ A. $2y - x$ cm

☐ B. $4y - 2x$ cm

☐ C. $2y$ cm

☐ D. x cm

23) What is the supplement of a 60° angle?

☐ A. 120°

☐ B. 130°

☐ C. 150°

☐ D. 180°

24) Solve the following inequality:

$$6x - 26 < 4x - 10 - 6x$$

☐ A. $x < 5$

☐ B. $x > -7$

☐ C. $x > 6$

☐ D. $x < 2$

25) Elsa answered 10 out of 50 questions on a test incorrectly. What percentage of the questions did she answer correctly?

☐ A. 20%

☐ B. 60%

☐ C. 75%

☐ D. 80%

17.2 Answer Keys

Section 1: Arithmetic Reasoning

1) A. 120
2) C. 80
3) B. 40%
4) B. $132
5) B. 59 seconds
6) A. 360 square feet
7) A. 1,000
8) D. 14
9) A. $29.40 an hour
10) B. 9
11) B. 11 inches
12) B. 20
13) D. 120 yards
14) A. 50 miles
15) C. 19
16) B. 100 hours
17) B. 9
18) C. 5%
19) B. 16.5 miles
20) D. 25 dimes
21) B. 360
22) C. 56
23) A. $p + 4$
24) A. $\frac{3}{4}$
25) B. 8.7%
26) D. -32
27) A. $150x + 18,000 \leq 25,000$
28) B. 10
29) D. $-\frac{7}{17}$
30) C. 77.5

Section 2: Mathematics Knowledge

1) A. $6x^2 + 12x - 48$
2) D. 0
3) B. 20 units
4) C. 98
5) A. 5
6) C. 72%
7) B. 55
8) D. 9
9) B. $y = x + 1$
10) C. 2
11) A. $4x^8$
12) C. $(-1, 2)$
13) D. It cannot be determined from the information given.
14) B. 5
15) A. 7
16) A. 25,200 square feet
17) C. 115 degrees
18) A. 12 cm, 16 cm
19) B. The y-intercept represents the starting height of 10 inches.
20) A. 54
21) C. 108
22) A. $2y - x$ cm
23) A. 120°
24) D. $x < 2$
25) D. 80%

17.3 Answers with Explanation

Answer Details — **Section 1: Arithmetic Reasoning**

1) Susan works 8 hours a day for 5 days a week, and this continues for 3 weeks. Therefore, the total hours worked are calculated as:

$$8 \text{ hours/day} \times 5 \text{ days/week} \times 3 \text{ weeks} = 120 \text{ hours}.$$

Thus, Susan has worked for 120 hours in total, making option A correct.

2) Set X includes numbers from 20 to 99 that are not in set Y. The count of these numbers is:

$$99 - 20 + 1 = 80.$$

Therefore, there are 80 integers in set X but not in set Y, making option C correct.

3) The number of female students is calculated as:

$$80 - 48 = 32.$$

The percentage of female students in the class is then:

$$\left(\frac{32}{80}\right) \times 100\% = 40\%.$$

Hence, 40% of the class is female, making option B correct.

4) The tax amount is 10% of $100, which is $10. The total bill becomes $100 + $10 = $110. The tip is 20% of $110, which is $22. The total expenditure is:

$$\$110 + \$22 = \$132.$$

Thus, Thomas's total expenditure is $132, making option B correct.

5) Lily's three fastest times are 55, 60, and 63 seconds. The average is calculated as:

$$\frac{55+60+63}{3} = \frac{178}{3} \approx 59.3 \text{ seconds.}$$

Therefore, the average of her three fastest times is approximately 59 seconds, making option B correct.

6) The area of the hall is calculated by multiplying its length and width:

$$\text{Area} = 18 \times 20 = 360 \text{ square feet.}$$

Therefore, 360 square feet of carpet is required, making option A correct.

7) To convert 2.54821 to 2548.21, the decimal point must move three places to the right:

$$2.54821 \times 1000 = 2548.21.$$

Thus, multiplying by $1,000$ achieves the desired number, making option A correct.

8) Among the options, 14 is the only number that does not evenly divide 100:

$$100 \div 14 \notin \mathbb{Z}.$$

Therefore, 14 is not a factor of 100, making option D correct.

9) John's new hourly rate after a 5% increase is calculated as follows:

$$\text{New rate} = \$28.00 + 5\% \times \$28.00 = \$28.00 + \$1.40 = \$29.40.$$

Therefore, John's new hourly rate is $29.40, making option A correct.

10) Let the number of photos Alice took be x. David took $3x$ photos, and Sarah took $x+18$ photos. Since

17.3 Answers with Explanation

David and Sarah took the same number of photos:

$$3x = x + 18 \rightarrow 2x = 18 \rightarrow x = 9.$$

Therefore, Alice took 9 photos, making option B correct.

11) Convert the heights to inches: Julia = $5 \times 12 + 9 = 69$ inches, Eva = $4 \times 12 + 10 = 58$ inches. The difference in height is:

$$69 - 58 = 11 \text{ inches}.$$

Thus, the height difference is 11 inches, making option B correct.

12) The average is calculated by adding the numbers and dividing by the count:

$$\text{Average} = \frac{18 + 26 + 15 + 21}{4} = \frac{80}{4} = 20.$$

Therefore, the average of the numbers is 20, making option B correct.

13) First, calculate the perimeter in feet: $2 \times (120 + 60) = 360$ feet. Convert to yards: $360 \div 3 = 120$ yards. Thus, the distance walked is a 120 yards, making option D correct.

14) The distance is calculated as speed times time:

$$\text{Distance} = 20 \text{ miles/hour} \times 2.5 \text{ hours} = 50 \text{ miles}.$$

Therefore, the total distance covered is 50 miles, making option A correct.

15) For the sum to be between 75 and 100, the average must be between 15 and 20.

$$\frac{75}{5} = 15, \quad \frac{100}{5} = 20.$$

The possible average values are 15, 16, 17, 18, 19, and 20. Option C, 19, is a valid average.

16) Calculate the time required:

$$\text{Time} = \frac{1,200 \text{ dishes}}{12 \text{ dishes/hour}} = 100 \text{ hours}.$$

Therefore, it will take 100 hours to prepare 1,200 dishes, making option B correct.

17) Calculate the average number of books organized per volunteer:

$$\text{Average} = \frac{100 - 10}{10} = \frac{90}{10} = 9.$$

Therefore, the average number of books organized per volunteer is 9, making option B correct.

18) Calculate the percentage increase:

$$\text{Increase} = \frac{\$8.40 - \$8.00}{\$8.00} \times 100\% = \frac{\$0.40}{\$8.00} \times 100\% = 5\%.$$

The raise represents a 5% increase, making option C correct.

19) Convert the map scale to actual distance:

$$\text{Distance} = 5.5 \text{ inches} \times 3 \text{ miles/inch} = 16.5 \text{ miles}.$$

The actual distance is 16.5 miles, making option B correct.

20) Calculate the total cost and convert to dimes:

$$\text{Total cost} = 50 + (2 \times 75) + (2 \times 25) = 250.$$

$$\text{Dimes required} = \frac{250}{10 \text{ per dime}} = 25 \text{ dimes}.$$

Thus, 25 dimes are needed, making option D correct.

21) To determine the number of rotations the second hand makes in 6 hours, we need to first calculate the number of minutes in 6 hours and then the number of rotations per minute.

There are 60 minutes in an hour. Therefore, in 6 hours, there are $6 \times 60 = 360$ minutes. Since the second

17.3 Answers with Explanation

hand completes one rotation (360 degrees) every minute, in 360 minutes, it will complete $360 \times 1 = 360$ rotations. The correct answer is B. 360.

22) Calculate the product of the square roots:

$$\sqrt{64} \times \sqrt{49} = 8 \times 7 = 56.$$

The product is 56, making option C correct.

23) Given 5 consecutive even integers, the difference between the least and greatest is 8:

$$\frac{p+(p+8)}{2} = \frac{2p+8}{2} = p+4.$$

The value of $\frac{p+q}{2}$ is $p+4$, making option A correct.

24) Calculate the additional sugar needed:

$$3\frac{1}{4} - 2\frac{1}{2} = 3.25 - 2.5 = 0.75 \text{ cups} = \frac{3}{4} \text{ cups}.$$

Therefore, $\frac{3}{4}$ cup of sugar is needed, making option A correct.

25) Convert the decimal to percent:

$$0.087 \times 100 = 8.7\%.$$

The percent equivalent is 8.7%, making option B correct.

26) Calculate using the order of operations:

$$-24 + 8 \times (-3) - \frac{6 + 18 \times (-3)}{3} = -24 - 24 - \frac{6-54}{3} = -48 + 16 = -32.$$

The result is -32, making option D correct.

27) Let x be the number of basketballs the team can purchase. According to the assumptions, we have:

$$150x + 18,000 \leq 25,000$$

Therefore, the inequality representing the number of basketballs is $150x + 18,000 \leq 25,000$, making option A correct.

28) Solve for x:
$$\frac{x+10}{x} = 2 \rightarrow x + 10 = 2x \rightarrow 10 = x.$$

The value of x is 10, making option B correct.

29) Substitute $m = 8$ and $n = -4$ into the expression and simplify:
$$\frac{7 - 9(4-4)}{4 \times 8 - 7(3+4)} = \frac{7}{32 - 49} = -\frac{7}{17}.$$

The value is $-\frac{7}{17}$, making option D correct.

30) Calculate the mean score: Total score for 3 tests $= 75 \times 3 = 225$,

$$\text{Total score for 4 tests} = 225 + 85 = 310, \rightarrow \text{Mean score} = \frac{310}{4} = 77.5.$$

Therefore, the mean score is 77.5, making option C correct.

17.3 Answers with Explanation

 **Section 2: Mathematics Knowledge**

1) Expand the expression:

$$(2x+8)(3x-6) = 6x^2 - 12x + 24x - 48 = 6x^2 + 12x - 48.$$

The expanded form is $6x^2 + 12x - 48$, making option A correct.

2) The repeating sequence in the decimal expansion of $\frac{1}{13}$ is "076923", which is 6 digits long. To find the 73rd digit, we need to determine which digit of the sequence corresponds to this position.

We divide 73 by 6 (the length of the repeating sequence). This gives:

$$73 \div 6 = 12 \text{ with a remainder of } 1.$$

The remainder tells us the position within the repeating sequence. Since the remainder is 1, the 73rd digit is the first digit of the repeating sequence. The first digit in the sequence "076923" is "0". The correct answer is D. 0.

3) The perimeter of the triangle is the sum of its sides:

$$5 + 7 + 8 = 20.$$

The perimeter is 20 units, making option B correct.

4) The greatest value of x divisible by 7 and less than 100 is 98 (since $98 \div 7 = 14$). Option C is correct.

5) Let the amount of juice produced by juicer 2 be x liters. Then juicer 1 produces $3x$ liters. Together they produce $4x$ liters, which equals 20 liters:

$$4x = 20 \;\rightarrow\; x = \frac{20}{4} = 5.$$

Juicer 2 produced 5 liters of juice, making option A correct.

6) Calculate the percentage of the essay Kylie completed yesterday and today: Yesterday's Completion =

10% of the essay. After yesterday, the remaining percentage of the essay:

$$\text{Remaining} = 100\% - 10\% = 90\%.$$

Today, Kylie completed 20% of the remaining 90%:

$$\text{Today's Completion} = 20\% \text{ of } 90\% = \frac{20}{100} \times 90\% = 18\%.$$

Total percentage completed so far:

$$\text{Total Completed} = 10\% + 18\% = 28\%.$$

Finally, calculate the percentage of the essay still unfinished:

$$\text{Unfinished} = 100\% - \text{Total Completed} = 100\% - 28\% = 72\%.$$

Therefore, 72% of the essay is still unfinished, making option C correct.

7) Set up the equation:

$$10 + 3x = 24 + 19 \rightarrow 3x = 33 \rightarrow x = 11.$$

Calculate $5x$:

$$5x = 5 \times 11 = 55.$$

The correct option is B, 55.

8) For every 4 green marbles, there are 3 red marbles. If there are 12 green marbles (which is 4×3), the number of red marbles is $3 \times 3 = 9$. Option D is correct.

9) Calculate the slope of line l using points $(5, 4)$ and $(2, 1)$:

$$m = \frac{4-1}{5-2} = \frac{3}{3} = 1.$$

Line l has a slope of 1. A parallel line must have the same slope. Option B, $y = x + 1$, has a slope of 1 and is

17.3 Answers with Explanation

parallel to line l.

10) List the odd factors of 90:
$$1, 3, 5, 9, 15, 45.$$

Only 9 and 15 are between 5 and 30. Thus, there are 2 such factors, making option C correct.

11) Simplify the expression:
$$y = (2x^4)^2 = 2^2 \times (x^4)^2 = 4 \times x^8 = 4x^8.$$

Therefore, the correct expression for y is $4x^8$, making option A correct.

12) Substitute the points into the equation $y = 4x + 5$:

$$\text{For } (-1, 2): y = 4(-1) + 5 = -4 + 5 = 1 \text{ (Does not match)}.$$

Point $(-1, 2)$ does not satisfy the equation, making option C correct.

13) Solve for S using the second equation:
$$4S + N = 20 \rightarrow 4S = 20 - N.$$

Substitute $N = 15 - M$ into the equation:
$$4S = 20 - (15 - M) \rightarrow 4S = 5 + M.$$

Without the value of M, we cannot determine the exact value of S. Therefore, option D is correct.

14) Calculate the distance using the distance formula:
$$d = \sqrt{(4-1)^2 + (6-2)^2} = \sqrt{3^2 + 4^2} = \sqrt{9 + 16} = \sqrt{25} = 5.$$

The distance between the points is 5, making option B correct.

15) Solve the quadratic equation:

$$x^2 - 49 = 0 \rightarrow x^2 = 49 \rightarrow x = \pm\sqrt{49} \rightarrow x = \pm 7.$$

Therefore, x could be 7, making option A correct.

16) The area of a rectangle is calculated by multiplying its length by its width:

$$\text{Area} = 140 \times 180 = 25,200 \text{ square feet.}$$

Thus, the area of the garden is $25,200$ square feet, making option A correct.

17) Since lines C and D are parallel, angle x and angle y are supplementary angles and hence:

$$x = 180 - 65 = 115 \text{ degrees.}$$

The value of angle x is 115 degrees, making option C correct.

18) Let the sides be $3x$ and $4x$. The perimeter of a parallelogram is twice the sum of its adjacent sides:

$$2(3x + 4x) = 56 \rightarrow 14x = 56 \rightarrow x = 4.$$

Therefore, the sides are $3 \times 4 = 12$ cm and $4 \times 4 = 16$ cm, making option A correct.

19) In the function of the tree's height over time, the y-intercept represents the initial height before growth. In this case, the tree starts at 10 inches, making option B correct.

20) Calculate the cube of 6 and then divide by 4:

$$\frac{6^3}{4} = \frac{216}{4} = 54.$$

One fourth of the cube of 6 is 54, making option A correct.

17.3 Answers with Explanation

21) Identifying prime numbers in the list and summing them:

$$17 + 23 + 31 + 37 = 108.$$

The sum of the prime numbers is 108, making option C correct.

22) Since P is the midpoint of QR, the length of PR is half of QR. Therefore:

$$PR = \frac{1}{2} \times (4y - 2x) = 2y - x \text{ cm}.$$

The length of segment PR is $2y - x$ cm, making option A correct.

23) The supplement of an angle is found by subtracting the angle from 180°:

$$180° - 60° = 120°.$$

The supplement of a 60° angle is 120°, making option A correct.

24) Solve the inequality:

$$6x - 26 < 4x - 10 - 6x \;\rightarrow\; 8x - 26 < -10 \;\rightarrow\; 8x < 16 \;\rightarrow\; x < 2.$$

Therefore, x is less than 2, making option D correct.

25) Calculate the number of questions answered correctly: $50 - 10 = 40$. The percentage of correct answers is:

$$\frac{40}{50} \times 100\% = 80\%.$$

Therefore, she answered 80% of the questions correctly, making option D correct.

Author's Final Note

I hope you enjoyed this book as much as I enjoyed writing it. I have tried to make it as easy to understand as possible. I have also tried to make it fun. I hope I have succeeded. If you have any suggestions for improvement, please let me know. I would love to hear from you.

The accuracy of examples and practice is very important to me. We have done our best. But I also expect that I have made some minor errors. Constant improvement is the name of the game. If you find any errors, please let me know. I will fix them in the next edition.

Your learning journey does not end here. I have written a series of books to help you learn math. Make sure you browse through them. I especially recommend workbooks and practice tests to help you prepare for your exams.

I also enjoy reading your reviews. If you have a moment, please leave a review on Amazon. It will help other students find this book.

If you have any questions or comments, please feel free to contact me at drNazari@effortlessmath.com.

And one last thing: Remember to use online resources for additional help. I recommend using the resources on `https://effortlessmath.com`. There are many great videos on YouTube.

Good luck with your studies!

Dr. Abolfazl Nazari

Made in the USA
Monee, IL
05 October 2024